Technicia
Technolo

This book covers the standard unit in Manufacturing Technology at Level 3 of the Technician Certificate in Mechanical and Production Engineering. The learning objective structure of the syllabus has been followed, and the content is also suitable for a wide range of TEC programmes having units in Workshop Technology. This book, together with *Technician Workshop Processes and Materials 1* and *Technician Manufacturing Technology 2*, is intended to provide all of the material required by the Technician student for a Technician Certificate programme which takes the study of Workshop Technology up to level 3.

C. R. Shotbolt is Principal Lecturer in Engineering Technology at Luton College of Higher Education. He is an established author and has written *Technician Workshop Processes and Materials 1, Technician Manufacturing Technology 2, Workshop Technology for Mechanical Engineering Technicians 1* and *2*, and is co-author of *Metrology for Engineers*. Mr Shotbolt is Chairman of the A5 Programme Committee set up by Bedford College of Higher Education, Dunstable College and Luton College of Higher Education to coordinate TEC programmes in Mechanical and Production Engineering in Bedfordshire.

Technician Manufacturing Technology 3

C. R. SHOTBOLT
CEng, FIProdE, MIMechE, MIQA, MASQC
Principal Lecturer in Engineering Technology
Luton College of Higher Education

CASSELL · LONDON

CASSELL LTD.
35 Red Lion Square, London WC1R 4SG
and at Sydney, Auckland, Toronto, Johannesburg,
an affiliate of Macmillan Publishing Co. Inc.,
New York
© Cassell Ltd. 1980

First published 1980

I.S.B.N. 0 304 30293 7

Printed and bound in Great Britain at
The Camelot Press Ltd, Southampton

Preface

This book is written specifically to cover the objectives set by the Technician Education Council (TEC) in unit U76/057, Manufacturing Technology 3, which is offered in the programmes for Mechanical and Production Engineering Technicians. The TEC suggests that the order in which these objectives are tackled may be varied as required and the author has adopted a slightly different sequence in places. To a small extent the objectives have been modified occasionally to enable this sequence to be achieved but, in one way or another, all of the TEC learning objectives are covered either in the text or in questions which can be answered after studying the text or by carrying out the laboratory work which is explained in the text. The unit, being at level 3, is necessarily an essential unit only in those programmes leading to a Higher Certificate having a production engineering content but it is likely to be offered as an optional unit in most other certificate programmes, as in many fields it is considered that all engineers should have a thorough grounding in manufacturing methods. To all who use this book it is hoped that they will find it of assistance in achieving their objectives.

As with the works at levels 1 and 2 which precede this book the TEC objectives are stated at the beginning of each section, and at the end of each chapter there is a series of questions of the type the student may face in assessments to see if the objectives set in the unit have been achieved. Some are short answer questions which have been designed to see whether the student can recall facts. Others ask the student to explain why things are done in a certain way. These are designed to check that the student understands the topic, while others are asked for which the student will not find a definite answer in this work. These are asked to probe the ability of the student to apply the principles outlined in the text.

When writing *Technician Manufacturing Technology 2*, the work preceding this, the author realised that some students would not proceed beyond level 2. Had the author rigidly adhered to the objectives of the level 2 unit some topics would have been left incomplete. Therefore, for students who stopped at level 2 in this subject the author included in that book some work from level 3. This has resulted in work being repeated in this book which has already appeared in the preceding work. In each case it has been considered to a greater depth and will, in any case, avoid reference back.

Where called for, reference has been made to the specifications of the British Standard's Institute but in very few cases have extracts from the tables of the Standards been reproduced. Instead the author has tried to explain the principles involved in the Standards so that the student understands what he is doing when using them. In the units on General and Communication Studies the student should have learned how to obtain information from library sources and, in any case, in industry he will have to refer to either British or company standards. It is better therefore that the student makes such reference while still a student when errors can be corrected at no cost.

Safety Notes have again been used, either at the appropriate points in the text or as separate sections, where that is considered more suitable. Particular mention is made of first aid in the chapter on welding and to the regulations governing the use of abrasive wheels although, as in the case of British Standards, attempts have been made to explain principles rather than reproduce first aid manuals or the statutory instruments to which reference has been made.

Acknowledgements have been made throughout the text to firms and organisations who have so willingly made information available. Colleagues at Luton College of Higher Education have also been most helpful; in particular Mr. W. Vann has always been happy to give advice and information when asked, and encouragement when he thought it necessary.

Finally, again, mention must be made of the ladies who typed the text from the author's handwritten script and, bearing in mind that much of this work came from previous books by the author, to the staff at Cassells who managed to piece together the typescript, illustrations and pages cut out of other books, into the volume which you, the reader, have before you now.

It is the author's hope that all who read this work will find it interesting and be successful. If in some small way your progress is advanced by the use of this book then writing it has been worthwhile.

Luton 1980   C.R.S.

Contents

1 Welding processes

General Objective: *The student should be able to recognise the principal welding processes.*

INTRODUCTION

The joining of metals by welding has been practised by man for about three thousand years. In those early days all welding was done by blacksmiths and was only practicable with iron. The parts to be welded were heated to about 1000°C until they were plastic and then hammered together. At these temperatures the iron oxide on the surface of the iron is fluid and the hammering was done in such a manner that the oxide was squeezed out of the weld, giving metal-to-metal contact.

Later as man's technology developed, methods became available of heating metals locally until they fused and the two parts which were being welded flowed into each other while molten, solidifying into a continuous joint. Losses and reinforcement were made up by the use of a filler rod of similar composition to the metals being joined. Thus developed the modern welding process.

Note that, unlike a soldered or brazed joint, a welded joint has the following characteristics:

1. A continuous joint is made, there being no foreign material 'sticking' the two parts together.
2. The continuity extends to the grain structure of the weld which, ideally, should be indistinguishable from that of the joined parts.
3. If a filler rod is used it simply makes good any losses and builds up the weld. It does not stick the parts together.

It follows that two basic methods of welding are available:

(a) *Pressure welding*, in which the parts are heated to a plastic condition and the weld is formed by mechanical pressure.

(b) *Fusion welding*, in which the parts are melted and run into each other, giving a continuous joint.

Variations available in these basic methods of welding are largely in the methods of heating used, fig. 1.1 showing those to be discussed in this work. It must be realised that these by no means exhaust the methods available. Modern techniques which will not be discussed here include arc-image welding, laser-beam welding, electron-beam welding, cold pressure welding, ultrasonic welding, explosive welding and many others.

METHODS OF FUSION WELDING

To produce a weld by fusing the two parts together requires an extremely intense and localised source of heat, so that the parts are melted locally and are little affected outside the weld area. Such intense high-temperature sources of heat can be obtained by the combustion of suitable gases or by an electric arc.

Specific Objective: *The student should be able to describe, with the aid of sketches, the principles of oxy-acetylene welding.*

GAS WELDING

Burning a fuel gas with air, as in a simple gas blow torch, will not give high enough temperatures for welding. Consequently the gas is burned with oxygen, the most common fuel gases being acetylene, hydrogen and propane, with flame temperatures as follows:

Oxygen – acetylene	3250°C
Oxygen – hydrogen	2800°C
Oxygen – propane	3100°C

OXY-ACETYLENE WELDING

Two systems are available for oxy-acetylene welding, (a) *low pressure* and (b) *high pressure.*

The low pressure system is generally used where the welding process is carried out at fixed points on a production line and large quantities of acetylene are required. The acetylene is produced in a generator and piped to the work stations, a special injector type of blowpipe being required.

The high-pressure system is used in general engineering, maintenance and garage work. It has the great advantages that it is portable, requires no power supplies and the capital cost of equipment is relatively low. The equipment required is as follows (see fig. 1.2).

(a) *Oxygen cylinder*
The oxygen is supplied at high pressure, approximately 120 atmospheres in a steel cylinder fitted with a high-pressure valve. Oxygen cylinders are painted BLACK.

(b) *Acetylene cylinder*
Acetylene becomes unstable if compressed to a greater pressure than 2 atmospheres and is then liable to explode. The gas may be dissolved safely at higher pressures in liquid acetone, which will dissolve 25 times its own volume of acetylene for each atmosphere of pressure. The gas is therefore dissolved at about 15 atmospheres pressure and a cylinder when full contains $15 \times 25 = 375$ times its own volume of gas. Acetylene cylinders are painted RED and have a flat end.

(c) *Pressure regulators*
Before the gases are fed to the blowpipe their pressures must be reduced by a pressure regulator. When the cylinder stop-valve is turned on, gas is supplied to the underside of a spring-loaded diaphragm. The gas pressure, if in excess of the spring pressure, holds the valve closed. If the pressure screw is now turned until the spring pressure exceeds the gas pressure, the diaphragm deflects, opens the valve and releases gas, whose pressure depends upon the amount by which the spring is compressed. Points to note are:

(i) Oxygen regulators have right-hand threads with plain nuts and acetylene regulators have left-hand threads, with notched hexagon nuts so that they cannot be confused.
(ii) Oxygen regulators should never be greased. Grease in contact with high-pressure oxygen is liable to cause an explosion.
(iii) The regulator is *closed* by *unscrewing* the regulating screw.
(iv) The regulators have two gauges; the high-pressure gauge shows the quantity (pressure) of gas remaining in the cylinder, while the low-pressure gauge shows the output pressure from the regulator.

(d) *The Blowpipe*
The gas blowpipe has two connections, one for the outlet union of each regulator. Each screw thread has the same hand as the corresponding cylinder and regulator so that the connections cannot be confused. The oxygen hose is black and the acetylene hose maroon. The welding nozzles of the blowpipes are interchangeable, varying in diameter from 1.0 mm for thin sheet work to 4.0 mm for heavy duty work. Gas is supplied from both regulators at the same pressure, and the adjustments on the blowpipe control the quantity of each gas supplied. Apart from mixing the gases, the blowpipe also contains a means of preventing flashback of the flame up the supply hoses.

To light the blowpipe, both regulators are adjusted to the same pressure, depending upon

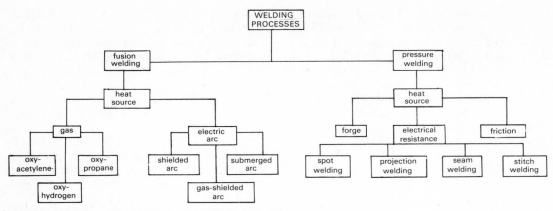

Fig. 1.1 Summary of welding processes discussed in the text.

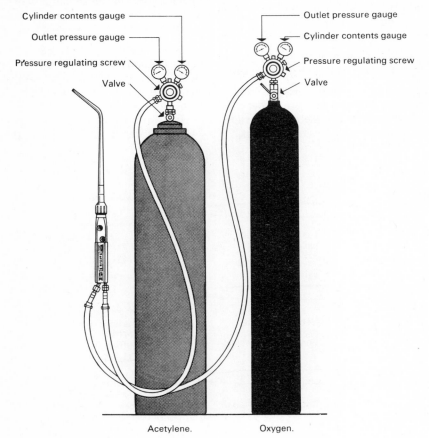

Acetylene.
(cylinder painted maroon)

Oxygen.
(cylinder painted black)

Fig. 1.2. High pressure welding outfit. (Reproduced by courtesy of British Oxygen Co. Ltd.) *Note*: Latest Continental practice is to use *blue* cylinders for oxygen.

the tip size being used. This varies from 13 kN/m² for the smallest tip to 50 kN/m² for the large tips. The acetylene adjustment on the blow-torch is opened first and the flame started with a spark lighter. This flame is simply one of acetylene burning in air and is bright yellow and long, and gives off black smuts. The oxygen adjustment is now opened gradually and as the quantity of oxygen increases the flame shortens and takes on a typical blue colour.

FLAME TYPES

Specific Objective: *The student should be able to state the three types of flame used in welding and the purpose of leftward and rightward methods of welding.*

By suitably adjusting the oxygen control on the blowpipe, three types of flame can be produced:

   (i) Oxidising flame – excess oxygen
  (ii) Carburising flame – excess acetylene
 (iii) Neutral flame – complete combustion with the minimum necessary oxygen.

Fig. 1.3 shows these flames diagrammatically. The correct adjustment for a neutral flame is attained when the haze due to excess acetylene just disappears. If the oxygen supply is increased, the cone shortens and the flame is oxidising in nature. If the oxygen supply is reduced, a feather of acetylene becomes more prominent and a carburising flame is produced.

For most purposes a neutral flame is desirable. If a carburising flame is used when welding steel the carbon content of the weld becomes greater than that of the parent metal and the weld becomes brittle.

A carburising flame is used for depositing Stellite, i.e., in hard-facing work, while an oxidising flame is used when welding alloys containing zinc, e.g., brass.

WELDING TECHNIQUES

Two methods are employed for producing a weld by gas welding. These are illustrated in fig 1.4(a) and (b) and are called *leftward* and *rightward* welding.

Leftward, or forward, welding is carried out with the torch in the right hand, the movement

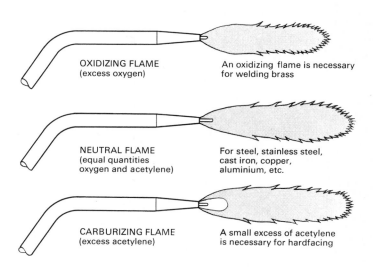

OXIDIZING FLAME
(excess oxygen)

An oxidizing flame is necessary for welding brass

NEUTRAL FLAME
(equal quantities oxygen and acetylene)

For steel, stainless steel, cast iron, copper, aluminium, etc.

CARBURIZING FLAME
(excess acetylene)

A small excess of acetylene is necessary for hardfacing

Fig. 1.3 Welding flame conditions. (Reproduced by courtesy of British Oxygen Co. Ltd.)

12

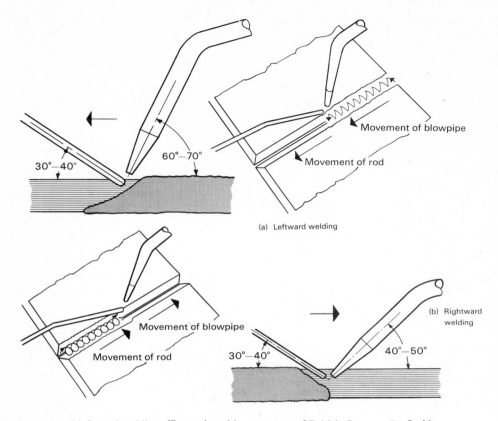

Fig. 1.4 Rightward and leftward welding. (Reproduced by courtesy of British Oxygen Co. Ltd.)

being from right to left, with the flame pointing into the unwelded portion of the work. This method gives better visibility for the operator but is limited to work of 4–5 mm thickness or less.

Rightward, or backward, welding uses the opposite technique, with the flame playing over the weld already produced. This reduces the cooling rate and has an annealing effect, improving the ductility of the weld. The method is used for thicker materials.

Fig. 1.5 relates the thickness to the edge preparation, filler rod diameter, technique to be used and the rate of welding in millimetres per minute for various thicknesses of metal.

FILLER RODS, FLAMES AND FLUXES

Specific Objective: *The student should be able to state the types and forms of supply of the filler metals and fluxes used in gas welding.*

It has already been made clear that in the welding process the filler rod is not used to 'stick' the two parts together – it is fused into the parent material to form a continuous joint which, ideally, should be indistinguishable from the parent metal. It follows that, in most cases, the filler rod is similar in composition to the parent metal. In certain cases it is necessary for

13

the rod to be slightly different to compensate for losses in the parent metal caused by the welding process itself.

Conversely, the welding process may cause harmful inclusions to be produced in the weld which could cause weakness and fluxes are used to prevent this. It follows that the flux itself must not react with the welded material at high temperatures but simply perform the function for which it is intended, i.e., to dissolve the oxide and float to the surface of the weld. Fluxes may be used as a dry powder but this may be blown clear of the weld area by the flame and therefore in gas welding the fluxes are often made into a paste, by mixing with water, and painted on the prepared joint before welding commences.

In general it is best to follow manufacturers' instructions in the choice of both welding rod and flux for a given operation. The following notes give some details of requirements for the welding of different materials and precautions which should be observed in the process.

(a) Mild steel

Filler rods, about 1 metre long, are supplied in varying thicknesses (see fig. 1.5). If the rods themselves are rusty, iron oxides will be included in the weld. For this reason the rods are copper-coated to prevent them rusting in storage before use.

(b) Medium and high carbon steels

Rods for welding these materials should be as nearly as possible the same composition as the parent metal. It is important that the carbon content is not lowered during welding by carbon being oxidised away and there must be no excess oxygen in the flame. A very slight excess of acetylene is therefore required – see fig. 1.3.

(c) Cast iron

This material is very weak in tension; if part of a casting is heated locally to a very high temperature for welding it will expand and cause tensile loads in other areas of the material. This can easily cause fractures and therefore iron castings should be preheated before welding.

During welding, cast iron does not become as fluid as it does during casting, but is welded at a pasty stage, a shorter, thicker filler rod being used than for steel welding. This filler rod is puddled in the weld to build up the material. The filler rod material is rich in silicon to prevent the formation of cementite which would give a hard white cast iron weld. The silicon promotes the formation of pearlite and flake graphite, to give a grey rather than hard white iron weld.

Flux is necessary to cause oxides and impurities to come to the surface of the weld, the flux coating being chipped away when the weld has cooled. The flux used is borax – the filler rod is heated, and dipped in the powder to become coated.

Welding of castings is normally carried out for repair, rather than for fabrication purposes, and the so-called *bronze welding* process is often used. This is not, strictly speaking, a welding process as the parent metal does not fuse – the two parts are joined by the filler rod, which is actually brass containing 60% Cu/40% Zn. This is obviously a much lower temperature process than true welding of cast iron and in most cases it can be carried out with no pre-heating. Thus, although the rod used is more expensive than the cast iron filler rod, the process is cheaper and the joint is just as strong as the cast iron weld.

(d) Stainless steels

Most stainless materials oxidise very easily, and it is the film of oxide, formed by atmospheric oxygen, which is stainless. Thus, when stainless steel is to be gas-welded, a flame with a slight excess of acetylene should be used to prevent oxidation of the alloying elements.

A flux is necessary to help prevent oxidation, and it is important that the underside of the joint is coated with flux to prevent atmospheric attack of the parts that cannot be seen.

The filler rod should be of similar composition to the parent metal and it is best to obtain detailed advice from the manufacturer as to the materials to be used.

(e) Aluminium and its alloys

Aluminium alloys are very good conductors of heat and have a low melting point. As the heat is rapidly conducted away from the weld area the material may suddenly collapse (the appearance of the metal gives no warning). For this reason the material must be well supported on metal blocks which conduct this excess heat away from the areas where it is not wanted.

Aluminium oxidises very easily and the oxide has a high melting point. The oxide must therefore be removed as it forms by using a flux which dissolves the aluminium oxide and floats to the surface of the weld. As the weld proceeds, the flux solidifies and protects the weld from further atmospheric attack. The flux used is supplied in powder form and may be made into a paste and painted over the weld area. A better method is to heat the rod and dip it in the powder, which adheres to it – and the powder is thus transferred to the weld.

The flame must never be oxidising, a neutral condition being preferred; but a very slight haze of acetylene is acceptable and preferable to the oxidising condition.

(f) Copper

Copper is produced by various methods, depending upon the purpose for which it is required. *High conductivity copper* is made by electrolytic deposition and is almost 100% pure. *Pitch copper*, produced by smelting, contains some oxygen and is not really suitable for welding – the hydrogen in the gas combines with the oxygen to produce porosity. *Deoxydised copper* has had the oxides removed during smelting by the addition of small amounts of phosphorous, and it is most suitable for welding. It follows that during welding the rod that is used should be of similar composition to the parent metal, to give an oxide-free weld.

A neutral flame should be used along with a suitable flux, based on borax or boric acid, and it is important that the central cone of the flame does not come in contact with the weld. This central cone contains a certain amount of unburnt fuel gas which, if it comes into contact with the hot copper, causes reactions which reduce the strength of the weld.

(g) Brass

Brass is an alloy of copper and zinc. Unfortunately, at temperatures suitable to melt the alloy, the zinc content tends to vaporise, leaving the weld deficient in zinc. For this reason the filler rod needs to be of a composition 2% to 3% richer in zinc than the parent metal. The zinc vapour produced reacts with atmospheric oxygen to produce quantities of zinc oxide powder which cloud the view of the operator and is unpleasant to inhale. Efficient ventilation is therefore necessary and a simple mask should be worn to cover the nose and mouth.

An oxidising flame tends to reduce the zinc losses and should always be used when welding brass. The flux used is a borax-based powder, similar to that used for the welding of copper.

FLAME CUTTING

Specific Objective: *The student should be able to describe, with the aid of sketches, the principles of oxy-acetylene cutting.*

If iron or steel is heated almost to melting temperature and a stream of oxygen is then played upon it, the metal is rapidly oxidised. The melting temperature of the iron oxide is less than that of the metal from which it is formed so the oxide, being in a liquid state, is blown away from the zone by the stream of oxygen being used.

It follows from this that the oxy-acetylene *cutting* blowpipe has two functions. It must:

(a) provide a suitable flame for preheating the metal and (b) provide a stream of oxygen for cutting.

The cutting torch therefore has three controls as shown in fig. 1.6. The acetylene control and the heating oxygen control provide the combustible gas mixture which burns at the annular jet surrounding the main central oxygen jet. The third control supplies pure oxygen to the central jet and is of the lever type, the supply of oxygen being cut off when the lever is depressed. The lever can be locked down by the cutting oxygen control.

To start the blowpipe, both oxygen valves should be closed and the acetylene valve opened, the gas jet issuing from the outer passage then being lit. The heating oxygen valve is opened and adjusted to give a suitable flame. The cutting jet is now started by opening the cutting oxygen valve, but the oxygen supply is cut off by depressing the central lever.

A guide clamped to the jet is rested on the work and, with the control lever depressed, the work is rapidly heated to white heat The lever is released and the stream of pure oxygen impinging on the work starts the cutting action. The torch can now be drawn slowly along the work to produce the cut.

With hand-held equipment, metal up to 600 mm thick can be cut. By machine cutting, metal up to 850 mm thick can be cut and greater thicknesses than this can be cut using an oxygen lance.

Profiles may be cut by hand, the torch being guided either to follow a chalk line or a suitably positioned metal guide. However, modern methods, particularly where a number of similar cut plates are required, use automatic profiling machines on which the torch is mounted on two slides, one moving at a predetermined speed along the work, while the other, moving at right angles, is controlled by a template to reproduce the template profile on the work. This slide is moved either by electrical or hydraulic means, similar to those used on copying lathes.

A disadvantage of this method is that the fixed slide moves at constant speed and it follows that the true cutting speed of the torch over the work is a combination of the speeds of the two slides and may thus vary, depending on the true direction of cut. More modern machines, where large amounts of plate are cut, as in shipyards, are tape controlled. In these cases the profile that is to be produced is programmed on a punched paper or magnetic tape which, through a tape reader and control devices, guides the torch to produce the desired profile.

Apart from the constant cutting speed, tape controlled cutting machines have other advantages in that the programmes can be easily changed to modify the profile and there are no bulky templates to store. Also the tapes take up much less room and are easier to handle.

The subject of tape control of machine tools will be dealt with in greater detail in *Technician Manufacturing Technology 4.*

HYDROGEN AS A FUEL GAS

As stated previously, if acetylene is compressed to above two atmospheres it becomes unstable and likely to explode. For this reason it cannot be used for underwater work, where a higher pressure is needed to force the gas out of the blowpipe against the water pressure. For underwater work, therefore, hydrogen is used. Although its flame temperature is lower than that of an oxy-acetylene flame, it can be used for normal cutting processes by divers working underwater.

The torch used for this work is similar to the normal cutting torch except that it has an additional annular jet or bell around the jet. This is supplied with compressed air to hold the water away from the flame and cutting area.

ELECTRIC-ARC WELDING

Specific Objective: *The student should be able to describe, with the aid of sketches, the principles of electric-arc welding.*

In this, as in gas welding, the parts to be welded are melted locally and filler rod is cast into the molten pool thus formed to build up and form an integral part of the weld. The localised heating is, however, produced by an electric arc struck either between the filler rod and the work, or between a separate electrode and the work.

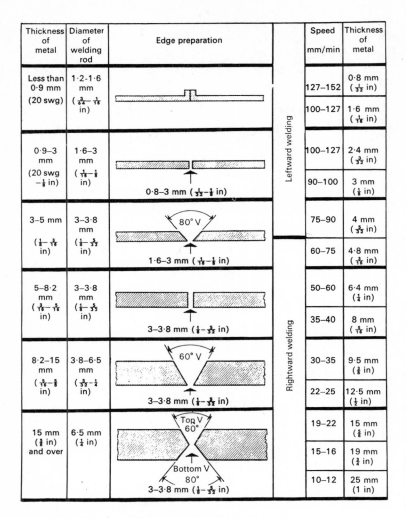

Thickness of metal	Diameter of welding rod	Edge preparation		Speed mm/min	Thickness of metal
Less than 0·9 mm (20 swg)	1·2–1·6 mm ($\frac{3}{64}$–$\frac{1}{16}$ in)		Leftward welding	127–152	0·8 mm ($\frac{1}{32}$ in)
				100–127	1·6 mm ($\frac{1}{16}$ in)
0·9–3 mm (20 swg –$\frac{1}{8}$ in)	1·6–3 mm ($\frac{1}{16}$–$\frac{1}{8}$ in)	0·8–3 mm ($\frac{1}{32}$–$\frac{1}{8}$ in)		100–127	2·4 mm ($\frac{3}{32}$ in)
				90–100	3 mm ($\frac{1}{8}$ in)
3–5 mm ($\frac{1}{8}$–$\frac{3}{16}$ in)	3–3·8 mm ($\frac{1}{8}$–$\frac{5}{32}$ in)	80° V 1·6–3 mm ($\frac{1}{16}$–$\frac{1}{8}$ in)		75–90	4 mm ($\frac{5}{32}$ in)
			Rightward welding	60–75	4·8 mm ($\frac{3}{16}$ in)
5–8·2 mm ($\frac{3}{16}$–$\frac{5}{16}$ in)	3–3·8 mm ($\frac{1}{8}$–$\frac{5}{32}$ in)	3–3·8 mm ($\frac{1}{8}$–$\frac{5}{32}$ in)		50–60	6·4 mm ($\frac{1}{4}$ in)
				35–40	8 mm ($\frac{5}{16}$ in)
8·2–15 mm ($\frac{5}{16}$–$\frac{5}{8}$ in)	3·8–6·5 mm ($\frac{5}{32}$–$\frac{1}{4}$ in)	60° V 3–3·8 mm ($\frac{1}{8}$–$\frac{5}{32}$ in)		30–35	9·5 mm ($\frac{3}{8}$ in)
				22–25	12·5 mm ($\frac{1}{2}$ in)
15 mm ($\frac{5}{8}$ in) and over	6·5 mm ($\frac{1}{4}$ in)	Top V 60° Bottom V 80° 3–3·8 mm ($\frac{1}{8}$–$\frac{5}{32}$ in)		19–22	15 mm ($\frac{5}{8}$ in)
				15–16	19 mm ($\frac{3}{4}$ in)
				10–12	25 mm (1 in)

Fig. 1.5 Edge preparations. (Reproduced by courtesy of British Oxygen Co. Ltd.)

The technique was first used in 1881, carbon electrodes being used to produce the arc, with a separate filler rod as in gas welding. In 1888 a bare steel wire was used as the electrode by a Russian, N. G. Slavianoff, the electrode also acting as the filler rod. This technique was used until the 1930's but it was difficult to strike and maintain the arc, and the weld was seriously contaminated by atmospheric oxygen and nitrogen. These difficulties have been overcome by covering the electrode with a thick coating of flux which shields the arc with large volumes of the gas given off, covers the weld with a layer of slag and controls the arc.

Later developments have been to flood the weld area with an inert gas to give gas-shielded arc welding, and to perform the weld under a deep layer of granulated flux, this method being known as submerged-arc welding.

Arc welding can be performed with either

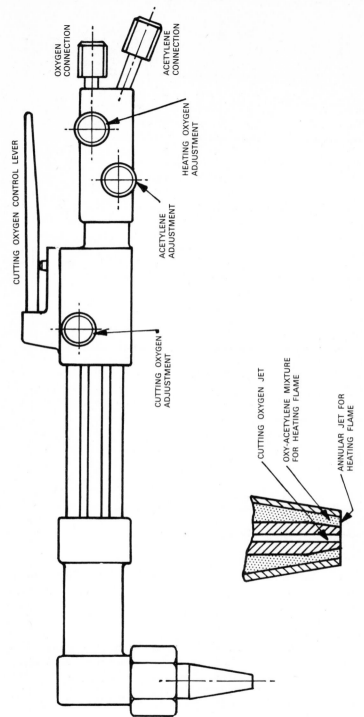

OXYGEN CONNECTION

ACETYLENE CONNECTION

HEATING OXYGEN ADJUSTMENT

CUTTING OXYGEN CONTROL LEVER

ACETYLENE ADJUSTMENT

CUTTING OXYGEN ADJUSTMENT

CUTTING OXYGEN JET

OXY-ACETYLENE MIXTURE FOR HEATING FLAME

ANNULAR JET FOR HEATING FLAME

Fig. 1.6 Flame cutting torch.

a.c. or d.c. electric power supply. Direct current requires the use of a motor-generator set or a transformer-rectifier, which are more complex electrically than the transformer used with alternating-current sets. For most purposes, therefore, a.c. welding sets are used but, as will be shown, direct current is used for certain gas-shielded processes.

METHODS OF ELECTRIC ARC WELDING[1]

(a) *Shielded-arc welding*

Specific Objective: *The student should be able to describe, with the aid of sketches, the principles of shielded-arc welding.*

The equipment used consists of a transformer to step-down the supply voltage and a regulator to allow the current to be adjusted to suit the requirements of the work. One connection from the regulator is made to the work and the other to the electrode holder, as shown in fig. 1.7.

[1] Types of filler rod and flux requirements are included in description of the processes.

The electrode is coated with a flux which, during welding, forms a cup in the end of the electrode and stabilises the arc. As the electrode is consumed, the coating gives off an inert gas which envelops the weld and prevents atmospheric attack. At the same time, other constituents form a slag which solidifies over the weld and gives a slow cooling rate and, again, prevention from atmospheric attack. If a weld requires more than one pass this slag must be chipped away, as it normally is from a finished weld. During welding, metallic particles are carried from the electrode into the weld as shown in fig. 1.8. This metal carry-over is due to the processes involved rather than gravity, which means that vertical and overhead welding can be readily accomplished.

The currents used vary up to 750A with a voltage from 50–100 V for a.c. welding, depending upon the diameter of the electrode and the work in hand.

To start the weld, the operator strikes the end of the electrode against the work rather like striking a match. The electrode is then withdrawn some 3 mm to 6 mm away from the work and, with the arc established, the rod is drawn slowly and evenly across the work to produce the weld.

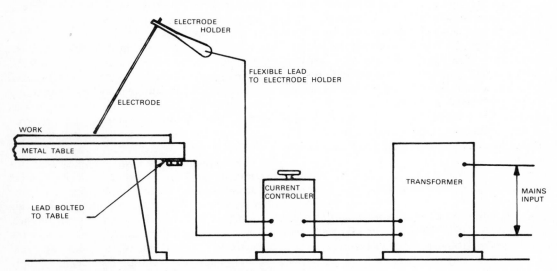

Fig. 1.7 Block diagram of electric arc welding set-up.

19

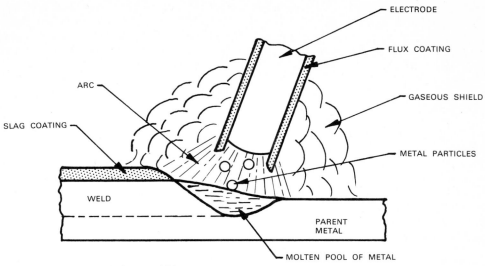

Fig. 1.8. Diagram of shielded arc weld in progress.

Due to the intensely localised heating of the parent metal the arc-welding technique can be used on castings without any need for pre-heating except for castings of intricate form. This is a considerable advantage over gas welding, where preheating of castings is normally essential.

Thin work, up to 6 mm thick, can be welded with one pass. For greater thicknesses, edge preparation is necessary as shown in fig. 1.9(a) and a sealing run is also needed if the work is done only from one side. For still greater thicknesses a number of runs are required on both sides, using different electrode thicknesses for the various runs. Fig. 1.9(b) shows a sequence of weld runs in a single-vee weld in a plate 25 mm thick. Comparison with fig. 1.9(c) shows the economy achieved by double-vee welds.

(b) *Gas-shielded arc welding*[1]

Specific Objective: *The student should be able to describe, with the aid of sketches, the principles of gas-shielded arc welding.*

[1] Often called TIG (Tungsten Inert Gas) welding.

Certain materials, notably those which oxidise readily, are extremely difficult to weld by conventional gas or electric-arc welding processes. To overcome the problems of attack by atmospheric oxygen the idea was conceived of using a non-consumable tungsten electrode to strike an arc and produce the melting temperatures required, and at the same time to flood the area of the weld with a continuous stream of inert gas.

Initially the gas used in America, where the process was developed, was helium and this welding method was called the *heli-arc* process. America was the only place where helium was readily available and in other areas argon was used. This gradually came to be preferred generally and the process is now called *argon-arc* welding.

The equipment is more complex than for conventional arc welding. Ideally, direct current would be used with the electrode made positive, as under these conditions any oxide formed is dispersed by the action of the arc. However, this tends to cause overheating of the electrode, which melts at the tip and contaminates the weld. In practice, a.c. supply is generally used, the positive half-cycle giving

20

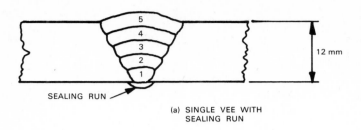

SEALING RUN

(a) SINGLE VEE WITH
SEALING RUN

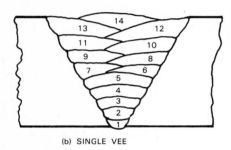

(b) SINGLE VEE

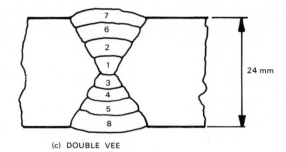

(c) DOUBLE VEE

Fig. 1.9. Sequence of runs for different conditions.

oxide removal and the cooler negative half-cycle keeping the electrode at a lower average temperature.

A high-frequency starter unit is required to initiate the arc since the electrode must not touch the work and thus become contaminated. Nor must it be touched by the filler rod.

A supply of argon is obviously required and, as this must flood the weld area, it is fed through the electrode holder around the electrode, an outline of the equipment being shown in fig. 1.10.

When the arc has been started it is held stationary for a few seconds until a molten pool of metal has formed. The end of the filler rod should be kept within the argon stream all the time, since constant withdrawal and replacement of the filler rod can disturb the argon flow and entrain air which will contaminate the weld. Correct choice of filler wire size is important and a leftward technique is used to flow argon ahead of the weld. When

the weld is complete the arc can be stopped but the flow of gas should be continued until the weld has cooled sufficiently for there to be no danger of oxidation of the weld area.

The process can be used to weld materials from 0.5 mm to 50 mm thick but high thermal conductivity may limit the maximum thickness that can be welded. Flux must not be used and the materials to be welded must be clean before welding starts. A feature of the process is the clean finish obtained.

(c) *Gas-shielded metal-arc welding*[1]

Although argon-arc welding has many advantages for some types of welding, particularly the absence of flux which might cause corrosion after welding, it was found to have limitations. A separate filler rod meant that the welder had two items to manipulate, while visibility and access to some joints tended

[1] Often called MIG (Metal Inert Gas) welding

21

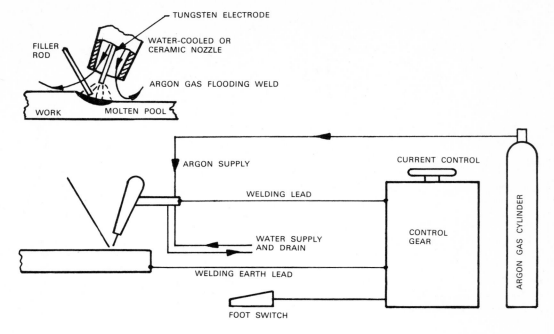

Fig. 1.10. Block diagram of gas-shielded tungsten-arc welding (argon arc or TIG welding).

to be poor. A later development was *gas-shielded metal-arc welding*, in which the filler rod becomes the electrode and is therefore consumable. This fact allows a direct-current system to be used, with the electrode as the positive pole, to give the desirable self-cleansing action mentioned earlier. The wire used is between 0.75 mm and 2 mm diameter and, being so thin, can be wound on a spool to give a continuous feed so that the operator does not have to keep stopping to change electrodes. A diagram of the set-up for this process appears in fig 1.11. Arrangements can be made for the arc length to be automatically controlled and the method therefore lends itself to complete automation.

(d) *CO$_2$ Arc welding of steel*
The gas-shielded processes discussed previously have considered only the inert gas.

argon as a shielding agent. Such a gas is necessary for shielding the welding of aluminium, stainless steel etc. but it is very expensive and the cost of the quantities required precludes its use for welding plain-carbon steel and cast iron, which can be successfully welded with the normal flux-coated electrode. However, the gas-shielded process does lend itself to automatic welding and for this reason the process has been modified to use carbon dioxide as the shielding gas for welding ferrous materials.

Flux-cored filler wires are used, varying in diameter from 1 mm to 4 mm. The wire is coiled on a spool and automatically fed through the welding 'torch' by rollers, the torch being water-cooled in heavy applications.

The speed of welding is very high, up to 500 mm/min., which largely eliminates distortion. Labour costs are claimed to be up to 50% less than for conventional manual arc welding.

22

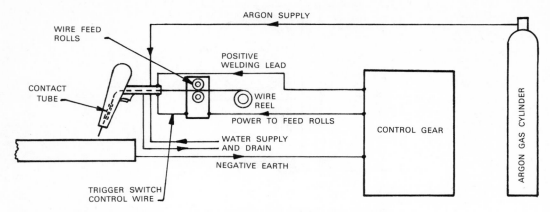

Fig. 1.11. Block of diagram of gas-shielded metal-arc welding (MIG welding).

(e) *Submerged-arc welding*

Specific Objective: *The student should be able to describe, with the aid of sketches, the principle of submerged-arc welding.*

The slag coating produced in manual arc welding is beneficial in that it protects the weld from atmospheric attack and also insulates it to give a reduced cooling speed. If these advantages are to be retained in automatic welding, problems arise in passing the current from the electrode guides, through the coating, to the electrode metal. These problems have been successfully overcome but it has been found better to feed the flux in powdered form from a hopper ahead of the electrode. The arc is therefore submerged[1] under a layer of flux which melts to form a slag over the molten pool and protects it.

A diagram of the process is shown in fig. 1.12. The wire is bare and needs no flux core as in CO_2 welding, but is copper-coated to prevent corrosion in storage and ensure good electrical contact with the electrode guide.

This method lends itself to downhand (horizontal) welding of mild and low-alloy steels. It is widely used in shipbuilding, for the

[1] Submerged-arc welding has nothing to do with underwater welding.

welding of pressure vessels, in pipe welding, etc., The welding head is moved over the work on a carriage, or the work is moved under a fixed welding head, the latter method generally being used in pipe welding, where the joint is revolved under the head.

(f) *Electro-slag welding*

As metal thicknesses increase, multi-run welds using automatic methods become uneconomical but if a large current is used to make the weld in one pass the molten pool tends to be uncontrollable and run ahead of the electrode. To overcome this, the plates to be welded are turned into the vertical position as shown in fig. 1.13 and water-cooled shoes are provided to contain the molten pool and prevent the slag from running away.

The wire electrode used passes through a bath of molten slag on top of the weld pool. This is not an arc welding process in the accepted sense. The heat is generated by the passage of the electric current through the molten slag, the slag bath also shielding the weld. As the electrode is consumed the depth of the molten pool increases and the water-cooled guides are raised. The process is similar to continuous casting.

The process is faster than submerged-arc welding for plates greater than 50 mm thick but the mechanical properties of the weld are slightly inferior.

23

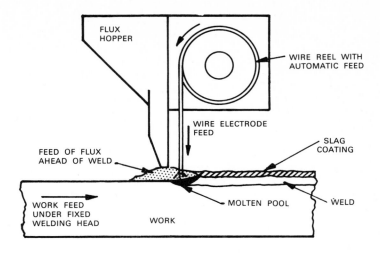

Fig. 1.12. Submerged-arc welding.

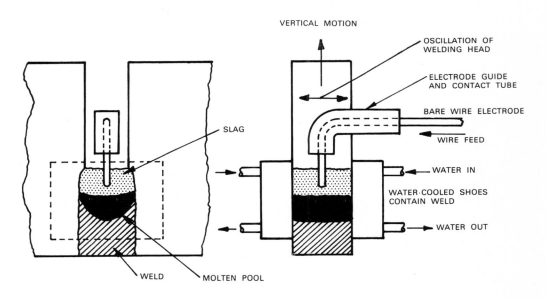

Fig. 1.13. Electro-slag welding.

COMPARISON OF FUSION WELDING PROCESSES

Specific Objective: *The student should compare welding processes, considering cost, speed, quality, ease of operation, and types of weld.*

The list below compares the processes discussed above and indicates briefly the applications of each process with the features peculiar to it.

1. *Oxy-acetylene welding*
Manual general-purpose process. Equipment of low capital cost and readily portable.

2. *Oxy-hydrogen welding*
As above but with lower flame temperature. Can be used under water.

3. *Shielded-arc process*
Faster than gas welding. Equipment more expensive and not so readily portable.

4. *Gas-shielded arc welding* (TIG welding)
Non-consumable tungsten electrode. Used for welding aluminium and magnesium alloys and stainless steel.

5. *Gas-shielded metal arc welding* (MIG welding)
Consumable electrode. Lends itself to automatic welding.

6. *CO_2 welding*
As for 5. Used on ferrous metals.

7. *Submerged-arc welding*
Automatic process for ferrous metals giving better slag protection than 6.

8. *Electro-slag welding*
As for 7 but can weld thicker sections.

DISTORTION AND STRESSES IN WELDS

If steel is loaded when cold, up to the elastic limit, it will deform until the load is removed, when it will return to its original shape. If the elastic limit is exceeded, the steel takes on a permanent set.

If the load is applied when the steel is at a temperature greater than 550°C, the steel does not exhibit any elasticity and any deformation due to the load will be permanent.

During welding, when the steel is heated to 1500–1600°C, it is therefore non-elastic. The non-uniform heating and cooling produce forces due to expansion and contraction, causing distortion which is permanent.

Furthermore, as the weld cools after solidification, further distortion occurs which, if restrained, causes residual stresses to be retained in the welded structure.

Consider a simple butt weld in thin material. The welded face of the material is heated to the plastic phase and has a much higher temperature than the underside. The forces of contraction are therefore much greater on the upper face and the plate bows upwards as shown in fig. 1.14. This is called longitudinal distortion.

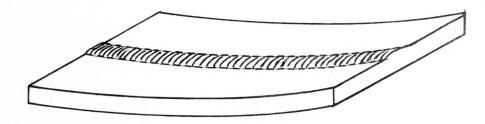

Fig. 1.14. Longitudinal distortion.

When a weld is being made where the edges have previously been prepared in the form of a vee, the inner faces of the vee expand inwards. When the weld is completed, a contraction in the same direction occurs, due to cooling of the deposited weld material. The distortion then takes place in an angular manner as shown in fig. 1.15(a). This can be overcome by presetting the parts at an estimated equal and opposite angle as shown in fig. 1.15(b), or by making a double-vee weld, where the contraction will be symmetrical.

Similarly the contraction of the weld fillets in a tee-joint can cause the cross-piece to bow as shown in fig. 1.16(a). Again this can be avoided by distorting the part as shown in fig. 1.16(b) prior to welding.

An alternative to presetting the parts is to

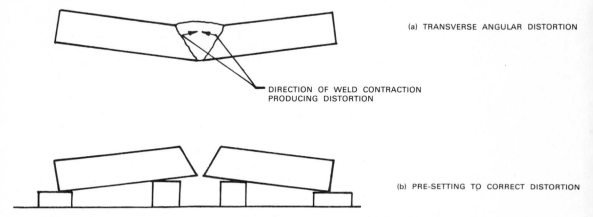

(a) TRANSVERSE ANGULAR DISTORTION

DIRECTION OF WELD CONTRACTION PRODUCING DISTORTION

(b) PRE-SETTING TO CORRECT DISTORTION

Fig. 1.15. (a) Transverse angular distortion, and (b) presetting to correct distortion.

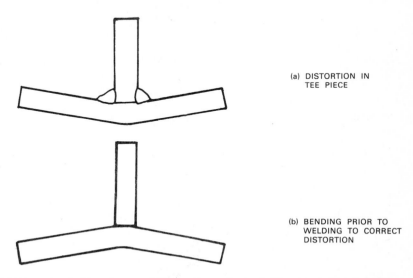

(a) DISTORTION IN TEE PIECE

(b) BENDING PRIOR TO WELDING TO CORRECT DISTORTION

Fig. 1.16. (a) Distortion in Tee piece, and (b) bending prior to welding to correct distortion.

26

restrain them by holding them in position in fixtures, or by the use of clamps, or by preliminary tack welding. This reduces distortion but the contraction forces are still present and, being resisted, set up residual stresses. Such stresses can be relieved by reheating the complete assembly, where feasible, or by locally reheating selected areas in the case of large structures.

If the heat input can be further localised by limiting the welding to a small length or area at a time, the amount of distortion will be reduced. Two methods of achieving this are (a) *step-back welding* and (b) *skip welding*.

In both cases the welding is done in short lengths along the work. In step-back welding, adjacent lengths of weld are made, each beginning where the next one will end, with the direction of deposit in opposition to the general direction of progression of the weld. This will be more clearly understood by referring to fig. 1.17(a). The technique used is rightward welding but the sections of weld are carried out in the numerical order shown, the arrows indicating the direction of deposition.

Skip welding is rather similar. A short length of weld is made at position 1, the next a short distance away at position 2 and so on as shown in fig. 1.17(b), the numbers again indicating the sequence and the arrows the direction.

Where a continuous weld is not essential intermittent lengths of weld can be used to advantage to reduce distortion, as shown in fig. 1.17(c).

PRESSURE WELDING

General Objective: *The student should be able to recognise the principal pressure welding processes.*

In the introduction to this Chapter it was mentioned that the earliest welding was performed by heating the metal, usually wrought iron, into the plastic temperature range and hammering the two parts together. The faces to be welded were usually curved, as shown in fig. 1.18, so that welding started at the centre and squeezed the oxide out.

This process is little used today and has been replaced by fusion welding of thick material, except in specialised cases such as flash butt welding, and by electrical-resistance welding such as spot, stitch, seam and projection welding for thinner materials. These latter processes have found extremely wide application in motor vehicle body manufacture, where the body is fabricated from sheet steel and the individual panels are spot welded to form the complex body structure of the modern automobile.

ELECTRICAL-RESISTANCE WELDING

Specific Objective: *The student should be able to describe, with the aid of sketches, the principles of electrical-resistance welding.*

An electrical-resistance weld is formed by passing an electric current through a localised area of the joint between the two parts to be welded. The resistance to the passage of the current at the joint between the surfaces raises the temperature there to a level which allows a pressure weld to be formed.

The localisation of the area of current passage is achieved by compressing the work between a pair of water-cooled copper electrodes as shown in fig. 1.19. Note that the requirements of the equipment for this type of weld are:

(a) A suitable electric current. This requires a transformer to step down the mains voltage and increase the current flowing.

(b) A means of applying pressure between the electrodes, by air or hydraulic pressure. Simple spring-loaded devices can be used but are not common nowadays.

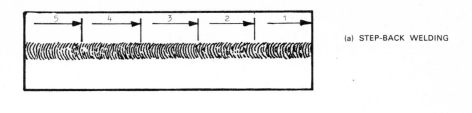

(a) STEP-BACK WELDING

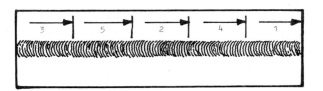

(b) SKIP WELDING

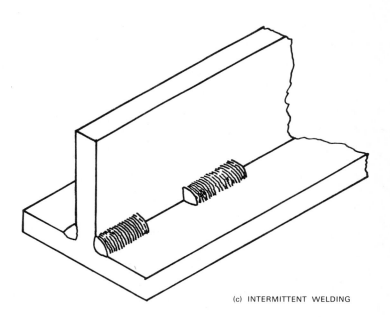

(c) INTERMITTENT WELDING

Fig. 1.17. (a) Step-back welding, (b) skip welding and (c) intermittent welding

28

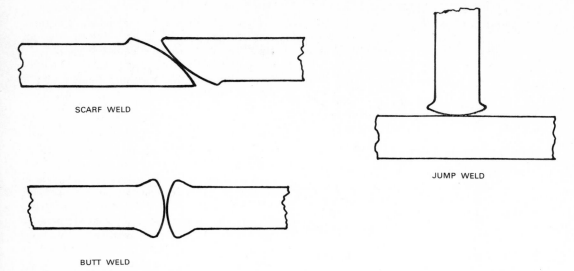

SCARF WELD

JUMP WELD

BUTT WELD

Fig. 1.18. Joint preparations for pressure welding.

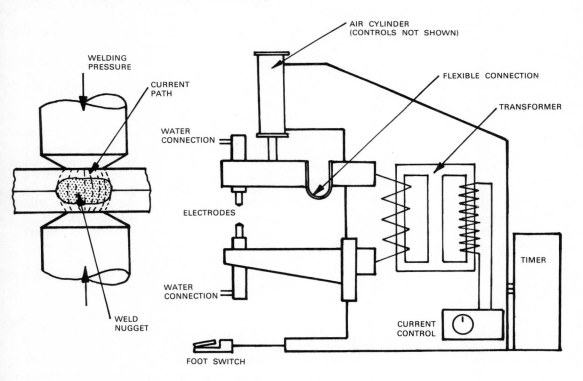

Fig. 1.19. Spot weld and block diagram of spot welding machine.

(c) A suitable timing device to give the correct sequence and length of events during welding.

The heat generated by the passage of an electric current is given by the expression:

$H = I^2 RT$, where H is the heat in joules
$\qquad\qquad$ I is the current in amperes
$\qquad\qquad$ R is the resistance in ohms
$\qquad\qquad$ T is the time in seconds.

Thus, the heat produced can be varied by:

(a) *Varying the current flow.* This is the least sensitive change which can be made, since the heat generated varies as the current squared. It is usually achieved by using a transformer whose tapping is changed by simply moving a control.

(b) *Varying the resistance.* The resistance to the flow of the current depends upon the pressure between the faces and the area through which the current flows. The area depends on the electrode size, which is decided by the metal thickness and cannot therefore be changed readily. The pressure used is that required to form the weld after heating has taken place and therefore cannot readily be altered.

(c) *Varying the time of current flow.* The heat generated depends directly upon the length of time through which the current flows, and this is the most sensitive control of the heat generated. The current time is measured in cycles of the a.c. supply at 50 Hz (hertz, formerly cycles per second) and modern timers can give a required time ± 0.25 Hz.

The foregoing is true for all types of resistance welding. Variations in the technique have been developed for different purposes as follows.

METHODS OF ELECTRICAL RESISTANCE WELDING

1. *Spot welding*

This process is illustrated in fig. 1.19, the two parts to be welded being pressed together between a pair of water-cooled electrodes, as shown. Usually, each spot weld is made individually, the spacing being determined by the operator although the approximate distance between spots should be given in the job specification to ensure adequate strength.

The electrode tip size varies, depending upon the thickness of the material, a guide to the tip diameter being:

$$D = \sqrt{t}$$

where D = electrode tip diameter in millimetres and t = thickness of material in millimetres with which the electrode is in contact.

A commonly-used design of electrode is shown in fig. 1.20, the tapered shank being used to hold the electrode in the electrode holder. An alternative type uses a threaded connection. The hole shown must be large enough to accept freely a tube from which cooling water can circulate.

An accepted figure for welding pressures is 70 MN/m² based on the electrode tip diameter. This pressure, often applied as an impact, or suddenly applied load, rapidly causes distortion of the electrode tip, which spreads, increases the contact area and reduces the intensity of pressure. It is therefore important to keep the electrode dressed to the correct form and diameter.

Ideally the weld should form a nugget as shown in fig. 1.19, the diameter of which is at least $0.8\sqrt{t}$, with the nugget equally distributed in the two materials. Faulty welds may be due to a number of causes, listed below. An indication of faulty welds is an enormous stream of sparks from the weld area. This usually indicates an excessive expulsion of metal from the weld area, giving a firework display which, although spectacular, often results in a weak weld of poor appearance.

Causes of faulty welds, in the order in which it is reasonable to check, are:

(a) Scaly or dirty material
(b) Electrodes improperly dressed
(c) Electrodes badly aligned
(d) Badly made parts. The electrode pressure is for welding, not correcting bad presswork.

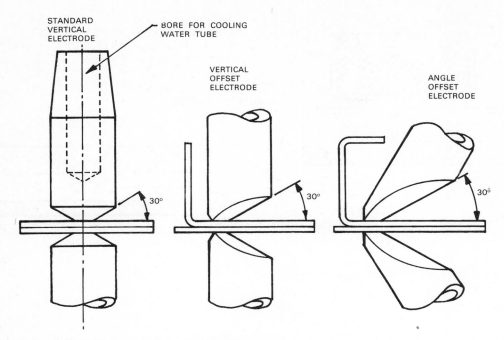

STANDARD
VERTICAL
ELECTRODE

BORE FOR COOLING
WATER TUBE

VERTICAL
OFFSET
ELECTRODE

ANGLE
OFFSET
ELECTRODE

30°   30°   30°

Fig. 1.20. Spot welding electrodes and their applications.

(e) Welding pressure incorrect (usually too low).

(f) Welding current incorrect (usually too high).

Spot welding is normally used for the production of individual welds between two sheet metal parts. If a solid part is to be welded to a sheet metal component or a number of welds are to be made simultaneously, the projection welding process is used.

2. *Projection welding*

A requirement of all methods of resistance welding is that the area through which the current passes and hence the heated area, is localised. In spot welding this is achieved by using small electrodes. If, however, a small area of one component is raised to form a projection a similar effect can be obtained. The electrodes are in this case shaped to suit the component and may assist in locating the parts.

If two sheet metal parts are welded in this manner, a number of welds can be formed simultaneously and the spacing of the welds can be closely controlled. The projections are produced by pressing during manufacture of the part and are of the form shown in fig. 1.21.

A typical case of welding a solid part to a sheet metal component occurs in the production of oil filters for automobiles, where an oil-tight connection is required. The boss is made of the form shown in fig. 1.22 and the annular projection collapses to form an oil-tight weld.

The production of long fluid-tight welds such as the seams of fuel tanks is not feasible by this method since the area of metal in contact is too great. For these purposes another method of resistance welding, seam welding, is employed.

3. *Seam welding*

To produce a water-tight seam by resistance welding, the two parts together are passed between a pair of discs which form the

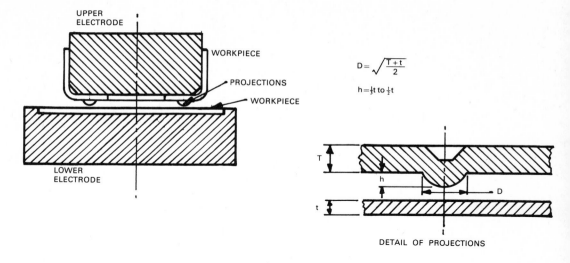

$$D = \sqrt{\frac{T+t}{2}}$$

$$h = \tfrac{1}{3}t \text{ to } \tfrac{1}{2}t$$

DETAIL OF PROJECTIONS

Fig. 1.21. Projection-welding sheet metal components. Note location in electrodes.

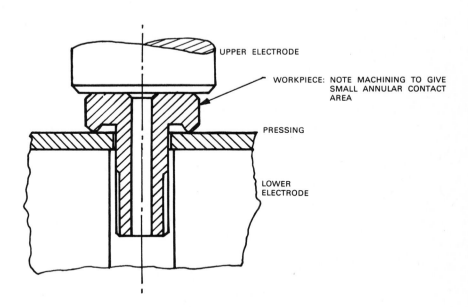

Fig. 1.22. Projection-welding machined boss to sheet metal pressing.

32

electrodes through which the current flows and between which the pressure is applied. As the discs rotate they roll the work between them and the current is passed in pulses to form a series of overlapping spot welds, as shown in fig. 1.23. Cooling is generally done by flooding the work from an external water supply.

If it is not necessary for the joint to be watertight, the speed of rotation of the discs can be increased so that the pulses, and hence the welds, are more widely spaced. In this case a series of non-overlapping spot welds is produced and the process is called *Stitch welding*.

A specialised form of seam welding is used in the production of butt-seam welded tubes. Flat strip is formed into tubular shape by passing it through rollers and formers. At the stage where it is completely formed and the edges touch, it passes under a pair of disc electrodes as shown in fig. 1.24, the current passing through the joint and raising the edges to the required welding temperature. The tube now passes through a further pair of rollers which exert the pressure to form the weld. The process is continuous and

tube is produced at speeds in excess of 20 m/min.

Methods of joining sheet metal, and solid pieces to sheet metal, have been discussed. A method of resistance-welding two solid parts together called flash butt welding is also available.

4. *Flash butt welding*

This process was developed during the 1939–1945 war when supplies of tungsten were short. To conserve this material, short lengths of high-speed steel, which contains tungsten, were welded to carbon steel shanks for cutting tools. Since then the process has become used for other purposes.

It is not, strictly speaking, a resistance-welding process in that the heating is produced by a form of arc. The parts to be welded are gripped in carefully aligned clamps, one fixed and one moving, as shown in fig. 1.25. The moving clamp is brought up until the parts touch and is then immediately moved back a short distance so that an arc is struck. This rapidly raises the temperature of the ends of the

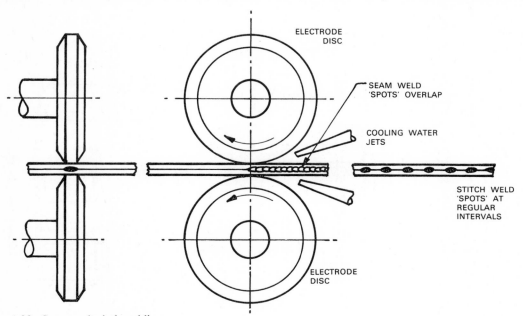

ELECTRODE DISC

SEAM WELD 'SPOTS' OVERLAP

COOLING WATER JETS

STITCH WELD 'SPOTS' AT REGULAR INTERVALS

ELECTRODE DISC

Fig. 1.23. Seam and stitch welding.

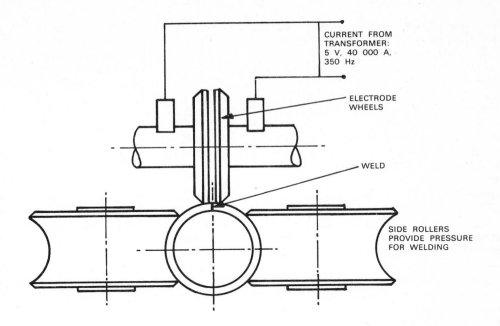

CURRENT FROM
TRANSFORMER:
5 V, 40 000 A,
350 Hz

ELECTRODE
WHEELS

WELD

SIDE ROLLERS
PROVIDE PRESSURE
FOR WELDING

Fig. 1.24. Butt seam welding of tube from strip.

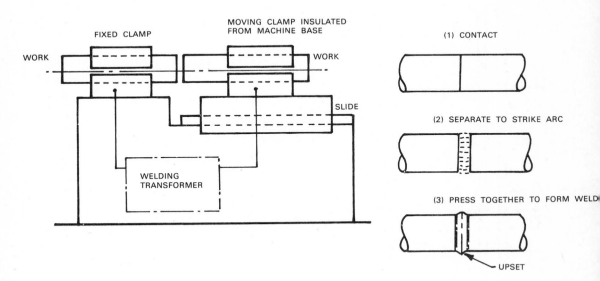

FIXED CLAMP

MOVING CLAMP INSULATED
FROM MACHINE BASE

WORK

WORK

SLIDE

WELDING
TRANSFORMER

(1) CONTACT

(2) SEPARATE TO STRIKE ARC

(3) PRESS TOGETHER TO FORM WELD

UPSET

Fig. 1.25. Diagram of flash butt welding machine and stages in the formation of weld.

34

parts and when they are hot enough for welding they are brought together again and pressure is exerted to form a weld.

The process is used today to save machining costs where a long rod with a head is required which it is not practicable to produce by cold-heading. An assembly with which the author was concerned is shown in fig. 1.26, the finished rod being afterwards projection-welded into a container. A feature of this component was that the upset produced by the flash butt weld was itself subsequently machined to form a shoulder on the part.

FRICTION WELDING

This is a recent development of pressure welding in which the required heat is generated by rubbing the two parts together. One component is held stationary while the other is rotated rapidly against it. When they are pressed together they quickly attain welding temperature, after which the rotating part is rapidly stopped and pressure applied to form the weld.

An interesting feature of the process is that it is self-cleansing and no oxidation of the surfaces can occur. This enables incompatible materials to be welded and the author has seen tubes of aluminium alloy and stainless steel welded together by this process. It can also be applied to thermo-plastic materials, in which industry it is known as *spin welding*.

The process is currently finding application in the automobile industry in the production of parts for rear axles, e.g., welding flanges to half–shafts. Control is very good and the process lends itself to complete automation.

METALLURGICAL EFFECTS OF WELDING

Specific Objective: *The student should be able to describe, with the aid of sketches, the effects of the welding process on the metal structure.*

During any welding process the metal in the weld itself, consisting of the parent metal and the deposited filler rod, has been melted, solidified and allowed to cool. During the welding process the parent metal in the vicinity of the weld has also been heated by conduction and cooled, the temperature reached being dependent upon the welding temperature, the speed of welding and the distance from the weld. The effects will also vary depending upon the conductivity of the material being welded, but as a general statement the above is true for all welding processes.

It follows that if a section is taken across a weld the temperature zones will be as shown in fig. 1.27. In the work on the heat treatment of steel in level 2 it was shown that the effect of heating to above the critical temperature

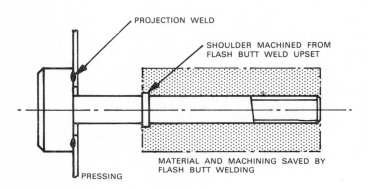

Fig. 1.26. Example of fabricated assembly using both projection and flash-butt welding.

35

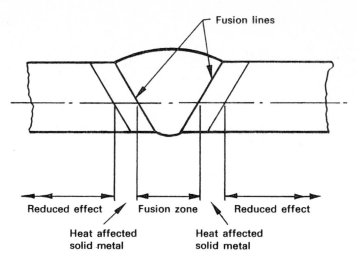

Fig. 1.27. Temperature zones across a weld.

followed by slow cooling is to produce grain growth and weakness. The higher the temperature, the greater will be the grain growth and therefore the closer to the weld, the larger will be the grain structure of the parent metal, the maximum grain size occurring in the parent metal immediately adjacent to the weld. This is where the weld will be weakest and where failure under load usually occurs. It is unusual for a good weld to fail in the weld itself – it is the parent metal next to the weld that fails.

In the weld itself the metal solidifies and forms metal crystals called *dendrites* which, much magnified, look rather like Christmas trees. These dendrites grow at right angles to the cooler parent metal giving, in single pass welds, a columnar structure as shown in fig. 1.28. This structure will also be coarse and weak, though not as weak as the parent metal immediately adjacent to the weld.

These structures can be improved by a normalising heat treatment as discussed in level 2[1].

In multi-pass welds the problem is not so critical, as each layer is reheated by the

---

[1] *Technician Manufacturing Technology 2* (Cassell) pp. 11–40.

succeeding passes, and this improves the structure of the lower layers and the material adjacent to the weld.

WELD TESTING

The only real way to determine the strength of a weld is to test it under load to destruction. By such destructive tests only a small sample can be tested and what is really being tested is the ability of the welder or the setting of the welding machine. Various methods of non-destructive testing are available which serve to disclose defects in the weld which affect its strength. Such methods of non-destructive testing apply equally to other work, as well as welded joints. Here we shall consider only the destructive tests of welds.

TESTING OF FUSION WELDS
For simple butt joints, the welder is required to produce a test piece. This is cut into sections across the weld and the different strips can then be submitted to different tests. These include tensile and impact tests, and a bend test using a rig whose basic dimensions are shown in fig. 1.29, in which, with the aid of a standard

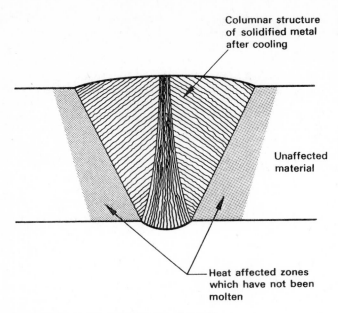

Fig. 1.28. Diagrammatic representation of structure of a weld.

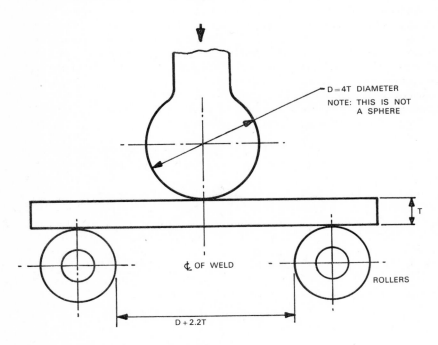

Fig. 1.29. Specification of bend test rig (B.S. 1295).

compression-testing machine, the test piece is bent through 180° into a 'U' shape, the surfaces of the weld having been machined level with the metal surface.

Such tests are covered by British Standards as follows:

BS 709: *Methods for testing fusion welds, welded joints and weld metal*
BS 1295: *Tests for use in the training of welders*
BS 2645: *Tests for use in the approval of welders.*

TESTING OF ELECTRICAL-RESISTANCE WELDS

Although attempts are being made to find non-destructive methods for spot welds, at present the only reliable test is to pull the welded joint apart. If the welds themselves separate at the joint, usually with a light load and a metallic 'ping', the weld is defective. If, however, the parent metal fails and leaves a 'slug' of the welded part as shown in fig. 1.30, the weld is considered good and the machine setting correct.

Such tests are usually carried out on a test specimen of similar material to that used for the components. A simple test is to hold one part in a vice and 'peel' the other part back with a pair of pliers until the weld fails. More reliable results are obtained by using a tensile testing machine and standard test pieces as shown in fig. 1.31.

Fig. 1.31(a), showing the shear-test specimen, is self-explanatory. Fig. 1.31(b) shows a cross-tension test piece, the two parts of which are bolted to suitable 'T' pieces held in the jaws of a tensile testing machine. A similar test may also be made on a 'U' tension test piece as shown in fig. 1.31(c).

Apart from these tests, impact and fatigue tests are also made on spot welds.

SAFETY NOTES
Specific Objective: The student should be able to state the sources of danger in welding and specify the necessary precautions and treatment.

Welding involves the manipulation of molten metal which has been raised to melting temperature by the use of either combustible gases or electricity. It follows that there are three obvious sources of danger: burns, explosions and electric shocks. Apart from these, other hazards are not so obvious and all must be considered if accidents are to be avoided.

1. BURNS
Burns may be caused by molten metal, flames or hot metal.

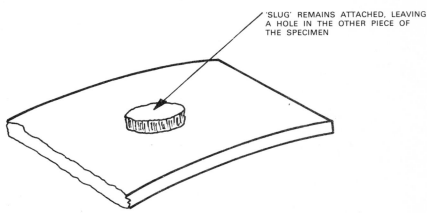

'SLUG' REMAINS ATTACHED, LEAVING A HOLE IN THE OTHER PIECE OF THE SPECIMEN

Fig. 1.30. Diagram of a good spot weld after test.

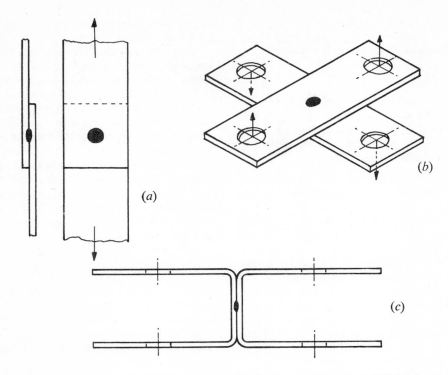

Fig. 1.31. Test pieces for testing spot welds. (a) Shear test. (b) Cross-tension test. (c) 'U' tension test.

(a) Molten metal

The heating processes used in welding are designed to give a very localised hot zone and the amount of molten metal in a weld at any time is very small. The danger of this running off to fall on the operator's foot is therefore low, the greatest danger being from globules expelled from the weld area as high temperature sparks. The greatest danger from these is that they may ignite clothing or enter the eye.

ALWAYS wear protective aprons over overalls and ALWAYS wear the correct goggles or head shields.

NEVER weld in oily overalls or clothes on which inflammable liquid has been spilt.

(b) Flames

Any flame which will melt steel will fry flesh to a crisp. As long as the operator is holding a lighted welding torch he can control it. When he puts it down the flame is almost invisible and at its most dangerous.

ALWAYS put the torch out when not actually welding.

NEVER act the fool with a welding torch or with anybody using one.

(c) Hot metal

The trouble with hot metal is that below about 500°C it looks the same as cold metal. Large work which cannot be moved should be clearly marked 'hot' until it has cooled off. Smaller, more portable pieces should be set aside to cool in a section of the workshop used and clearly marked for that purpose.

(d) Electrical burns

Where an electric current enters the body a

burn occurs which will often be deeper than it appears. If caused by an arc the sides of the burn will be cauterised and difficult to heal.

TREATMENT OF BURNS

The aim of first aid treatment of burns is to reduce the local effects, relieve pain, prevent infection of the burned area, replace fluid loss and, if the burns are severe, get the patient to hospital as quickly as possible.

1. To relieve pain, place the affected part under cold running water or immerse in cool water for at least ten minutes.
2. Remove rings, belts, boots, watches, and anything else which may constrict the burn, before it starts to swell.
3. Only remove clothing if soaked in boiling water. Dry, burned clothing has already been sterilised by the heat and should not be removed.
4. Cover the injury with a clean dry dressing.
5. Give small cold drinks (water) at frequent intervals.
6. Arrange for removal to hospital.

NEVER apply lotions, ointments or oil dressings.

NEVER prick blisters, breathe on or touch the burned area – this only increases the risk of infection.

2. EXPLOSIONS

Explosions are usually thought to be caused by the mixing and igniting of two substances which have very high speeds of flame travel (the 'explosiveness' of explosives is measured in terms of their speed of flame propagation). However, another type of explosion is caused by sudden expansion of gas under pressure. Both types can occur in a welding shop and should be guarded against.

(a) *Explosions due to combustion*

If the acetylene valve of a welding torch is opened and the torch is not lit or an acetylene regulator is opened to atmosphere, the gas released will form an explosive mixture with the atmospheric oxygen if the release takes place in a confined space.

Note also that an explosion can be caused *whatever* method of welding is used, when a tank or compartment which has contained fuel is being repaired. Such tanks must always be flushed out completely to remove residual fuel gas or vapour before welding commences.

ALWAYS ensure that the torch is ignited if the acetylene is turned on.

ALWAYS use a simple and quick means of ignition. The time taken fiddling with matches or a cigarette lighter can allow an explosive mixture to form. A small pilot flame on a fixed welding bench or a reliable spark igniter with a portable outfit are best.

Acetylene becomes unstable and spontaneously combustible at pressures greater than two atmospheres. In the bottle this is overcome by dissolving the gas in acetone (see p. 10), but in underwater welding the water pressure must be overcome.

NEVER use acetylene in a situation where it must escape against pressure – hydrogen is the fuel gas for such work.

(b) *Explosions caused by expansion*

Oxygen is stored at about 120 atmospheres. If an oxygen cylinder ruptures, it literally goes off like a bomb and causes a great deal of damage. The author has never seen this happen but he has seen a photograph of the effects of an oxygen cylinder exploding and it had taken the side out of the ground floor of a building, including the structural pillars near the explosion, and the whole building was in imminent danger of collapse. The author did once see a lorry on fire which included in its load a pair of welding bottles. It was interesting to note that it was the *oxygen* cylinder which occupied most of the firemen's attention – not the acetylene cylinder.

ALWAYS handle oxygen bottles carefully – particularly full ones.

3. ELECTRIC SHOCKS

When the human body is subject to an electric

shock it is the current value (amperage) which does the damage. From Ohm's Law:

$$I = \frac{E}{R},$$

it can be seen that, for a given voltage E, the lower the value of the resistance R, the higher will be the current I. If, therefore, R is the resistance of the body, most of which is produced by the skin, it follows that if the skin is wet and its resistance thereby lowered, the value of I will be raised and the likelihood of a severe shock is much greater. Equally, if E is high, I will be high and there will be a similar risk of a severe shock. This is not always true. If the internal resistance of the source is high as in the case of the high voltage off the sparking plugs of an engine, the shock may not be harmful but it is a sound general rule to treat all sources of electricity with respect — particularly those carrying high voltages.

By the nature of the set-up electric welding processes are well earthed but wherever mains electricity is being used, in these cases on the mains side of the equipment, there is danger.

NEVER touch the mains equipment when wet.

ALWAYS get repairs done by a qualified electrician.

TREATMENT FOR ELECTRIC SHOCK
1. Send for a qualified first aider and ambulance.
2. Before attempting any treatment make sure that the equipment is switched off.
3. If the equipment cannot immediately be isolated, only try to move the patient by pushing clear with **dry** insulating material **while standing on dry insulating material**.
4. If breathing has stopped, apply artificial respiration.
5. Treat any burns.
6. Keep warm and give **NOTHING** by mouth (not even tea and even less, alcohol).

4. ARC EYES
An electric arc of the type used in arc welding is an intense source of ultra-violet rays. These can be dangerous if viewed directly and can produce an extremely painful condition called *arc eyes*, particularly if the eyes are unshielded when the arc is struck. If a flash is observed directly it may be several hours before the full effects are felt. There will be a feeling that the eyes are full of sand and there will be a tendency to avoid light. Avoid flashes by keeping away from the process unless personally involved and, if involved, use only the correct type of glass shield. Special glass is required and ordinary smoked or tinted glass is not satisfactory.

TREATMENT FOR ARC EYES
1. Bathe the eyes in a solution of boracic acid to which a little glycerine has been added.
2. Lie in a dark room with a damp cloth over the eyes until the effects have passed.

Of necessity, in a book such as this, the treatments suggested for various types of injury are given briefly. Reference has been made to the book *First Aid*, the authorised manual of the St. John Ambulance Association and Brigade. Although much of the manual refers to topics outside the scope of this book the author feels that a study of its contents is of value to anybody, no matter what their walk of life.

SUMMARY

Welding in its earliest form was carried out by a combination of heat and pressure. It was used for artistic work, such as the great iron pillar at Delhi, and for the manufacture and repair of tools and implements. Later, fusion welding was developed when flame temperatures produced by the combustion of fuel gases with oxygen gave an intense source of heat suitable for the local melting of metals. These methods were, and still are, largely used in the repair and jobbing shop fields. The application of electricity to produce the heat required by an electric arc gave faster welding speeds and welding moved into the field of production, particularly in the areas of heavy engineering such as shipbuilding, bridge building and other structural work. At the same time the wheel turned full circle with the introduction of pressure welding as a volume-production process, using electrical resistance to give the required source of heat.

In all these fields there has been a continuous development, particularly in automatic methods, which have been facilitated largely by developments in electronics. Flame-cutting is carried out on numerically-controlled machines using identical control methods to those used on computer-controlled milling machines.

Thus the technician and the technician engineer working in the field of welding also requires a considerable knowledge of electricity and electronics. Although specialist knowledge in this field will largely come from his electrical technician colleague it behoves him to familiarise himself with the processes available if not their electrical details.

QUESTIONS

1. Explain the differences between soldering, brazing and welding in terms of
   (a) temperatures required;
   (b) strength of joints;
   (c) form of final joints.

2. List three gases, used in conjunction with oxygen, to produce welding flames, stating the temperatures attained and the type of work for which each is used.

3. Sketch three different types of welding flame, labelling the different zones of combustion and state one use for each type of flame.

4. Sketch the edge preparation for the welding of:
   (a) mild steel sheet 0.5 mm thick;
   (b) mild steel 2.0 mm thick;
   (c) medium carbon steel 6 mm thick;
   (d) mild steel plate 10 mm thick.

5. Explain with the aid of diagrams the meaning of the terms *leftward* and *rightward* welding. State the different effects of the two methods on the weld and which would be used for each of the welds in Question 4.

6. (a) What is the function of a filler rod in welding?
   (b) Under what circumstances may the composition of the filler rod differ from that of the parent metal.
   (c) State one case in which the filler rod is of the same composition as the parent metal and one case in which the compositions differ.

7. (a) What is the function of a flux in welding?
   (b) How must the density of the molten flux compare with that of the molten metal?
   (c) Name one flux commonly used in welding and the two metals on which it is used.

8. Explain with the aid of diagrams the methods of electric arc welding known as:
   (a) shielded arc welding;
   (b) TIG welding;
   (c) MIG welding.
List the advantages and limitations of each method and name an application of each.

9. MIG and CO_2 welding both use a gas to shield the weld area from atmospheric attack. Compare the advantages and limitations of each process and state an application of each giving reasons for the choice of method in these applications.

10. In the oxy-acetylene cutting process the metal is not removed by melting. Is this statement correct? If so, explain how the metal is removed and sketch a torch suitable for cutting, relating the reasons for its form of construction to the above statement.

11. Both a.c. and d.c. supplies are used at the arc in the electric-arc welding process. List the advantages and limitations of each, stating one application of each.

12. Sketch a cross-section through a spot-weld and explain the conditions necessary for producing the weld.
   Make a line diagram of a spot welding

machine and explain how the conditions for producing the weld are achieved.

13. The heating effect of an electric current is given by

$$H = KI^2RT$$

where K is constant.
Identify the terms I, R and T, and explain how each may be changed in a spotwelding machine.

In a particular weld the pressure used is producing holes in the weld and must be reduced. What electrical effect would this reduction in pressure have and what changes may be necessary to compensate for the change?

14. What is meant by the terms *stitch welding* and *seam welding* in the context of resistance welding? Use diagrams to explain the process and the difference between them.

15. Flash-butt welding is not, strictly speaking, a resistance welding process. Justify this statement by explaining how the process is carried out and name an application of the process.

16. The difference between friction welding and resistance welding is in the heat source. Explain this statement and make a simple diagram showing the essentials of a machine for friction welding.

Friction welding has a peculiar advantage over all other types of welding. What is it?

17. If a good weld is tested to destruction, it is usually the metal around the weld which fails rather than the weld itself. Explain, in terms of heat treatment of steel, why this is so. Illustrate your answer with a diagram showing the different zones in and around the weld.

18. Select suitable methods of welding for the following types of work, in each case giving reasons for your choice and show edge preparations where necessary:
   (a) Welding the domed ends on a high pressure gas cylinder.
   (b) Welding the stainless steel inner container of a piece of hospital sterilising equipment.
   (c) Welding the half-shaft to the flange for a motor car rear axle (two methods).
   (d) Welding a thick-walled aluminium cylinder to contain gas under pressure.

19. Name two sources of danger in electric arc welding, other than electric shock and state the treatment for each.

20. An operator has sustained a severe electric shock, is unconscious and is still in contact with the electrical equipment. The body is blocking the way to the switch gear and the apparatus is still switched on.
List the actions to be taken in terms of:
   (a) isolating the body;
   (b) first aid after isolation;
   (c) action by another person at the same time as first aid is being given.

21. Name two sources of danger in oxy-acetylene welding.

A portable gas welding outfit is close to a fire in a workshop. Portable extinguishers and a wall-mounted fire hose are available. How should these be used pending the arrival of the fire brigade?

22. A welder has dropped a lighted oxy-acetylene torch. In falling, the flame has burned his leg, set his overalls on fire and, on hitting the floor, the blow torch has gone out. List the actions to be taken, in order, paying particular attention to the treatment of the burned leg.

2   Precision casting

General Objective: *The student should be able to identify the principles of die casting, shell moulding and investment casting.*

INTRODUCTION

In level 2[1] the casting process was discussed as a primary production process, sand casting being considered at some length and mention being made of other methods of manufacturing castings which included die-casting, shell moulding and investment casting.

A detailed study of sand casting is more properly the province of the foundry technician and here we shall consider it no further. The mechanical engineering technician is likely to require more knowledge of the other casting processes, which lend themselves to high-volume production of very accurate castings and these methods will now be considered further. Die-casting is essentially a process for casting alloys of low melting temperature. It follows that no costly melting equipment is necessary and indeed the melting equipment is an integral part of the machine in many cases. For this reason the process is often used in firms which do no other form of casting, a bay of one of the production areas being given over to a few die-casting machines. Of course, the bulk of die-castings is made by specialist firms but the mechanical engineering technician is quite likely to come into direct contact with the process. The treatment here of die-casting will therefore be rather more detailed.

[1] *Technician Manufacturing Technology 2* (Cassell), pp. 41–65.

DIE-CASTING

Specific Objective: *The student should be able to describe, with the aid of sketches, the principles of die-casting by gravity and pressure methods.*

Many components are required whose shape is such that they can most easily be produced by casting but whose strength requirements are so low that they can be made of a relatively weak, low melting-point alloy. Where such parts are required in quantity it is economical to make a permanent metal mould in which the metal is cast. The fact that the mould is permanent enables a high production rate to be achieved, for as soon as one casting is removed the mould is ready for the next.

The mould used in this process is made from a mating pair of *dies* and the process is therefore known as *die-casting*. Die-castings are made by three methods:

1. Gravity Die-Casting
2. Cold-chamber Die-Casting   } *Pressure*
3. Hot-Chamber Die-Casting   } *Die-Casting*

In all cases the high finish of the die gives a high surface finish to the casting. The accuracy of the die is similarly repeated in the casting, so that little final machining is normally required.

GRAVITY DIE-CASTING
This process is known in the U.S.A. as *permanent-mould casting*, which is possibly a better name since it tends to prevent the process from being considered as the poor relation of pressure die-casting. In the gravity process a permanent mould and core are used, usually made of cast iron, and the metal to be cast is

poured into the mould as it is for a sand casting. The only pressure involved is the hydrostatic pressure due to the head of molten metal.

Except for one-piece dies for very simple parts, gravity dies are usually made from a number of blocks which can be separated to remove the part. This also permits the die blocks to be removed in a set sequence so that the component is allowed to contract freely instead of being constrained by the die. For this reason, gravity die-casting is not so senstitive to changes in section as pressure die-casting. The thin sections contracting naturally, do not fail as they would if restrained during cooling. Further, the process is flexible and allows design changes to be made by changing one or two die blocks.

As gravity dies are freestanding, neither the product designer nor the die designer is restricted by having to design dies for a particular machine of given platen size, stroke and metal capacity.

Thus gravity die-casting is an extremely useful process in its own right, advantages being summarised as follows:

1. Wide size-range of castings possible, from as small as 50 g mass up to 100 kg.
2. Less sensitive to changes in section than other methods of die-casting.
3. Readily adaptable to design changes.
4. Design of work not restricted by machine considerations.
5. Low capital cost of equipment compared with pressure die-casting.

PRESSURE DIE-CASTING

Unlike gravity die-casting, in pressure die-casting the metal is forced into the mould under a positive pressure which is maintained until solidification has taken place. The essentials of the equipment required are as follows:

1. Split dies containing a negative cavity of the form to be produced.
2. A positive method of closing the dies, holding them closed under pressure and opening them when solidification has occurred.

3. A means of forcing the molten metal into the dies under pressure and maintaining the pressure until solidification has taken place.
4. A means of ejecting the cast component when solidified.

Apart from the actual process of casting, some specialised finishing and machining operations are necessary before the casting can be used, and must be considered.

A fundamental difference between pressure die-casting and gravity die-casting is that the dies are made of hardened steel and of as few parts as possible.

(a) The cold chamber process

The basic principle of the cold chamber process is shown in fig. 2.1. A measured quantity of molten metal is ladled into the pouring hole and the ram is moved forward under pressure to force the metal into the dies. After solidification the dies are opened and since they are usually designed so that the work tends to remain in the moving die, the work is ejected by the ejector pins.

Use of the cold-chamber process is largely restricted to the casting of those alloys which have higher melting temperatures, such as aluminium alloys which melt at approximately 660°C, compared with 420°C for the zinc-based die-casting alloys. In the cold-chamber process the dies, sprue and cylinder have time to cool between shots and do not attain the same temperature as the alloy. This prevents 'iron pick-up', a transfer of iron from the machine and die steels to the alloy, which can occur if steel is in continuous contact with aluminium, magnesium and copper-based alloys for long periods at high temperatures.

Generally, the higher the melting point of the alloy the shorter the die life, even in the cold-chamber process. For this reason the metal temperature should be controlled carefully to the lowest temperature at which it can be successfully cast. In the case of copper-based alloys, such as brass, with a melting temperature of the order of 900°C, die life was found to be extremely low. For this reason

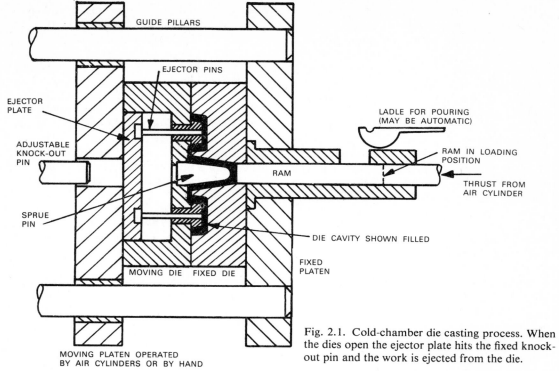

GUIDE PILLARS

EJECTOR PINS

EJECTOR PLATE

ADJUSTABLE KNOCK-OUT PIN

SPRUE PIN

LADLE FOR POURING (MAY BE AUTOMATIC)

RAM IN LOADING POSITION

RAM

THRUST FROM AIR CYLINDER

DIE CAVITY SHOWN FILLED

FIXED PLATEN

MOVING DIE   FIXED DIE

MOVING PLATEN OPERATED BY AIR CYLINDERS OR BY HAND

Fig. 2.1. Cold-chamber die casting process. When the dies open the ejector plate hits the fixed knock-out pin and the work is ejected from the die.

much of the die-casting of brass is now carried out with the metal in a pasty, or semi-liquid, condition at a much lower temperature.

The Polak machine used for die-casting brass in this manner has a vertical sleeve into which the semi-molten metal is placed. The ram now comes down and forces the metal into the die as shown in fig. 2.2. This inevitably leaves a residue of metal in the sleeve in the form of a slug which would prevent removal of the work from the dies, so the counter-plunger is forced up to shear off and eject this slug before the dies are opened and the part is ejected.

Generally, brass die-castings are cheaper than hot brass stampings but tend to be porous and cannot be made to the same density as hot stampings. Brass die-castings should therefore not be used to carry fluids under pressure.

(b) *The hot-chamber process*

Unlike the cold-chamber process, the hot-chamber process incorporates the metal-

melting equipment in the machine itself. As with the cold-chamber process, the moving die is mounted on a moving platen as shown in fig. 2.3. The fixed die and platen are attached to the melting pot, the connection to the sprue or die inlet being by a 'goose-neck', as shown. When the injection plunger is withdrawn molten metal flows into the goose-neck via a filling port. The dies are closed and the injection plunger is depressed, forcing the molton metal into the die cavity.

Usually the injection pressure of up to 140 MN/m^2 is created by hydraulic pressure. On large machines the die closure is performed by the ram of an air cylinder, the ram operating a toggle linkage so that the dies cannot open even if there is a failure in the pneumatic circuit. On smaller machines the die closure is performed by a hand-operated lever with a toggle action to ensure that the dies remain closed against the injection pressure.

The hot-chamber process is generally used

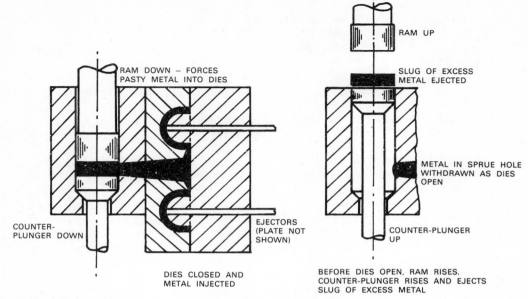

RAM DOWN – FORCES PASTY METAL INTO DIES

RAM UP

SLUG OF EXCESS METAL EJECTED

METAL IN SPRUE HOLE WITHDRAWN AS DIES OPEN

COUNTER-PLUNGER DOWN

EJECTORS (PLATE NOT SHOWN)

COUNTER-PLUNGER UP

DIES CLOSED AND METAL INJECTED

BEFORE DIES OPEN, RAM RISES. COUNTER-PLUNGER RISES AND EJECTS SLUG OF EXCESS METAL

Fig. 2.2. Polak cold chamber process for die-casting semi-molten metals.

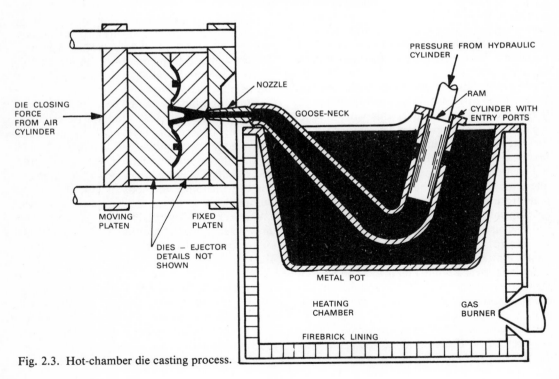

PRESSURE FROM HYDRAULIC CYLINDER

NOZZLE

GOOSE-NECK

RAM

CYLINDER WITH ENTRY PORTS

DIE CLOSING FORCE FROM AIR CYLINDER

MOVING PLATEN

FIXED PLATEN

DIES – EJECTOR DETAILS NOT SHOWN

METAL POT

HEATING CHAMBER

GAS BURNER

FIREBRICK LINING

Fig. 2.3. Hot-chamber die casting process.

47

for the lower melting temperature zinc-based alloys. Its design lends itself to automatic operation and it is faster than the cold-chamber process since ladling of the charge from a separate melting furnace is avoided.

SPECIAL FEATURES OF THE DIE-CASTING PROCESS

Specific Object: *The student should be able to state the special features, advantages and limitations, of the die-casting process*

Apart from the high speed of production, the die-casting process has certain other advantages which are of particular interest. It can produce castings of complex shape and with very thin walls. Thus weight can be kept to a minimum and by careful design ribs can be cast in which stiffen and strengthen the casting where necessary. The finish produced is almost a replica of that in the dies and in many cases a combination of good finish and high accuracy eliminates machining where it would be necessary if other methods of casting were used.

Screw threads and gears can readily be cast. If possible the die is designed with a split line on the diameter of the thread so that when the dies are opened the thread can be removed from the cavity without unscrewing. This method leaves a slight flash on the thread which has to be cleared by a machining operation. An alternative is to cast the part on a removable plate from which it is unscrewed while another is being cast.

The wall thickness of the casting can be from 1.5 mm to 5 mm and to a tolerance of about ± 0.08 mm. Cored holes require a draft taper of $1°$ for holes over 25 mm, falling to $\frac{1}{2}°$ on holes of 10 mm diameter or less. The tolerances on hole centres are of the order of ± 0.05 mm up to 40 mm, and ± 0.025 mm for each additional 25 mm above that if the cores are located in the same die block. If cores are located in different die blocks a greater tolerance is necessary.

USE OF INSERTS

Specific Objective: *The student should be able to give reasons for and examples of the use of inserts in die-castings.*

Although die-casting is a widely used high production process for the manufacture of castings it must be realised that the materials usually used — zinc-, aluminium- and magnesium-based alloys are generally weak and not greatly resistant to wear. A screw thread pulled down very tightly will easily strip and a highly loaded bearing wear rapidly. These problems can be overcome by the use of inserts. Instead of tapping a hole directly in the die-casting a brass insert can be used whose anchorage has a much greater shear area than the thread and can therefore withstand greater tightening loads. At the same time the thread in the insert material can withstand greater loads and the screw can be tightened to a higher torque without damage.

Similarly, although zinc-based die-casting alloys can give good bearing properties they are only suitable for lightly loaded bearings. Where higher bearing loads are required better bearing materials can be incorporated into a die-casting as an insert.

Studs and bushes can be cast into the work, the inserts usually being made of brass. They must of course be inserted into locations provided in the die while it is open and designed so that they remain anchored when the casting has solidified. Coarse knurling is often sufficient for this but other methods include the use of slotted-head screws, deliberately deformed components and slotted inserts, while bushes may be anchored by knurling or by milling flats on the outside diameter.

FINISHING DIE-CASTINGS
When the die-casting is ejected from the machine it is still attached to the sprue, runner and possibly a rim, used to avoid ejector pins operating on the cast surface. These attachments are generally broken off but there is

also a thin 'flash' around the joint line. This is removed by a flash-clipping operation in a simple press tool. An alternative method, where the profile is complex and a flash-clipping tool would consequently be expensive, is to run the components by hand around a routing tool running at a speed of about 30 000 rev/min. If the joint line runs across a flat face such as a flange, an abrasive belt with a backing plate is frequently used, as shown in fig 2.4. In both cases, only a light contact pressure is required.

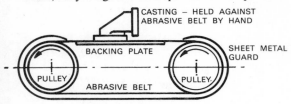

Fig. 2.4. Simple belt sanding machine for removing split line flash from die-castings.

Die-cast alloys are readily machineable and, although a feature of the die-casting process is that its accuracy obviates most machining, all the usual processes such as drilling, reaming,

tapping etc. can be carried out. An interesting technique is the finishing of round holes for bearings by a push broach of the type shown in fig. 2.5. In this example, the cutting teeth gradually increase in diameter from 12.25 mm to 12.50 mm, there being ten such teeth. The last four cutting teeth are all of 12.50 mm diameter to allow for regrinding, and the final four 'teeth' do not have cutting edges but are rounded, polished and made oversize by 0.005 mm as shown. These finish the hole by 'swaging' and in so doing impart a very good surface finish with fine surface cracks due to the work hardening which is caused. These cracks help to retain lubricant and help to provide an excellent bearing surface. The operation is very simple and is carried out by hand in a mandrel press, the broach dropping straight through the fixture into a suitable container under the press.

Many die-cast components are used for decorative brightware, particularly in the motor industry. Zinc-based die-castings lend themselves readily to nickel and chromium plating. The part is first polished, cleaned and then copper plated. This enables a coat of nickel

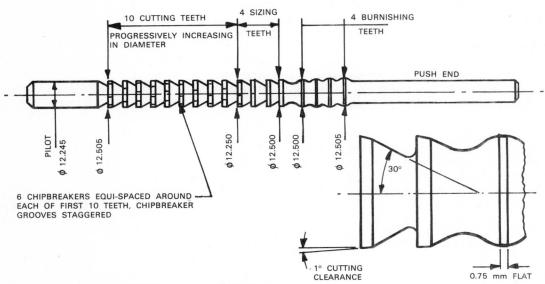

Fig. 2.5. Push broach for finishing round holes in die-castings.

49

plate to be deposited on the copper, after which the thin final layer of chromium plating is added and the final polishing is carried out.

Aluminium alloys do not lend themselves to finishing by electro-deposition but are more readily finished by anodising. Unlike plating, where the work is made the cathode in an electrolytic bath and the current flow is *towards* it, the aluminium alloy is made the anode and the current flows away from it. This produces a hard oxide surface which can be dyed to a required colour and polished. The anodised surface will not tarnish and is used for various purposes, from finishing items of domestic equipment to badges and buttons for the armed forces.

SAFETY NOTE

A cautionary note on the die-casting process should be added here. If for some reason the dies do not close properly the molten metal will be shot out with great force and can cause bad burns if it hits a person even some distance from the machine. To ensure that the dies close properly they should be cleaned with an air blast between each shot so that any foreign matter is removed. The machine should be guarded from floor to ceiling – Hardboard on a light wooden frame is adequate – and the guards should extend for some distance on either side of the split line of the dies. The operator may have to move around inside the barriers but he should wear safety glasses and suitable protective clothing and boots.

Die-casting is a precision casting process limited to the production of castings in low-melting point alloys. Obviously processes are required which will produce castings of equal precision in materials whose strength is greater and whose melting point is higher than the zinc-based and aluminium alloys discussed up to now.

SHELL MOULDING

Specific Objective: *The student should be able to describe, with the aid of sketches, the principles of the shell moulding process.*

The author understands that this process was developed in Germany during the 1939–1945 war. It enables a precision sandmould to be made from a metal pattern in a few minutes and in such a way that, apart from good accuracy (± 0.1 mm in 100 mm), a reasonable surface finish is produced.

Briefly the process consists of the following stages:

(a) Making the metal pattern
(b) Making the shell moulds
(c) Pouring the casting.

The pattern plates are made in pairs, corresponding to the two halves of a split pattern used for sand casting, and are made from metal, usually mild steel, as shown in fig. 2.6. Note that one plate has depressions to produce location bosses on the mould and the other plate has bosses to produce the mating depressions on the other half of the mould. Draft is required on the pattern plates to enable the finished moulds to be stripped from them and core prints are also necessary in this case. Patterns to form the runners are included but risers may not be necessary, depending upon the casting. This is because the finished moulds are porous (it is possible to blow cigarette smoke through them) and entrapped air can escape through the sides of the moulds.

Contraction must be allowed for in two ways. The pattern will be heated to about 450°C and will therefore expand, giving a larger mould than would be obtained from a cold pattern, and the molten metal, when cast, will shrink on cooling.

The mould is made from a mixture of sand and uncured synthetic resin. It is contained in a dump box as shown in fig. 2.7, and the heated pattern is clamped to the open end of the box, which is then inverted. This dumps the sand/resin mixture on the pattern, the heat from

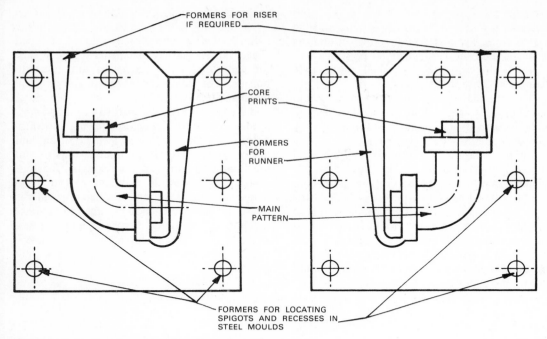

FORMERS FOR RISER
IF REQUIRED

CORE
PRINTS

FORMERS
FOR
RUNNER

MAIN
PATTERN

FORMERS FOR LOCATING
SPIGOTS AND RECESSES IN
STEEL MOULDS

Fig. 2.6. Pair of pattern plates for a flanged pipe bend.

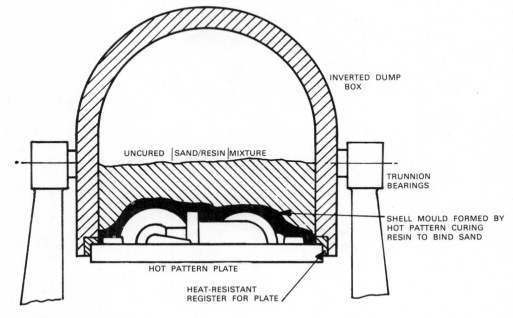

INVERTED DUMP
BOX

UNCURED SAND/RESIN MIXTURE

TRUNNION
BEARINGS

SHELL MOULD FORMED BY
HOT PATTERN CURING
RESIN TO BIND SAND

HOT PATTERN PLATE

HEAT-RESISTANT
REGISTER FOR PLATE

Fig. 2.7. Making a shell mould.

51

which cures the resin evenly to a depth depending on the pattern temperature and the curing time. The box is turned upright and the uncured resin and sand fall off the pattern which can then be removed. The shell mould thus formed is removed from the pattern.

The moulds are usually made in pairs and can be easily stored until required. They are light in weight and small in volume, so they occupy a minimum of floor space.

Before pouring, the moulds are clamped together as shown in fig. 2.8 by simple 'G' clamps. In practice, the shape of the component is not outlined as clearly on the outside of the mould as appears in fig 2.8. The cured sand/resin mixture tends to have a blurred outline as indicated in fig. 2.7. Accurate mould alignment is ensured by the registers provided on the patterns and reproduced on the mould

shells. If necessary, the mould can be reinforced by a backing of dry moulding sand but the metal is often poured directly into the clamped shells. No special arrangements are necessary for venting because the shells are naturally porous. The castings produced are of good finish and high accuracy, are remarkably sound and are free from scabs, porosity or blowholes.

Shell moulding can be used for high melting-point materials and due to their accuracy and finish, the castings produced require no machining on non-critical working surfaces. A typical component is a domestic water tap, whose hexagonal section on which a spanner is fitted is accurate enough not to require machining. Another and more complex casting is the finned cylinder for a two-stroke motor cycle engine, often made by the shell-moulding process.

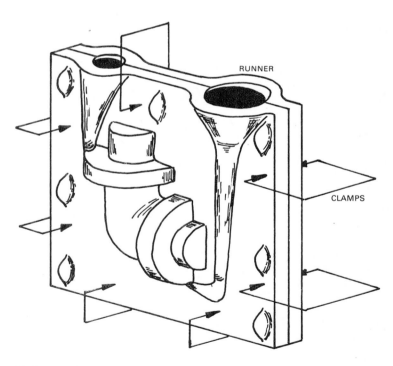

2.8. Shell mould after pouring.

INVESTMENT CASTING (LOST WAX PROCESS)

Specific Objective: *The student should be able to describe, with the aid of sketches, the principles of the investment casting process*

When the gas turbine engine was developed it was found necessary to produce alloys for the turbine blades which were heat resistant and also resistant to the phenomenon known as 'creep', i.e., a gradual increase in length under a steady load at high temperatures. It is unfortunate that alloys with these properties are extremely difficult to machine and casting techniques had to be found to produce blades of such precision that an absolute minimum of machining was necessary. Ironically, one of the oldest casting processes known was adopted – the lost wax process – which has been used for centuries to cast statues, emblems, religious figures etc. throughout the East. It is particularly suitable for an intricate casting whose pattern would be difficult to remove from the mould.

A replica of the casting is made or carved in wax, which is covered in layers of fireclay and allowed to dry. The fireclay is then baked and the wax pattern melts and runs out, leaving an internal hollow space which is a replica of the pattern inside the fireclay mould. Molten metal is poured in and solidifies on cooling to produce the casting required. The casting is removed by breaking off the surrounding fireclay.

Modern techniques use an accurate permanent metal mould in which a replica of the component is cast in wax, complete with runner sections. The resulting wax patterns are fixed to a central wax sprue by a hot soldering iron to form a 'Christmas Tree' of patterns. The whole assembly is then sprayed with a fine refractory material which dries to form a coating adhering to the wax. Successive coats are given until a sufficient body of ceramic has built up and dried. The whole assembly is now baked to harden and strengthen the ceramic mould, during which the wax replica melts and is lost – hence the 'lost wax' process – leaving a mould of high accuracy and surface finish into which metal of high melting-point can be cast.

The mould produced by the above method is rather fragile and a more robust mould is made by the *flask method*. The process is similar to that described above but before the mould is baked it is placed in a flask as shown in fig. 2.9 and a coarse refractory slurry is poured and vibrated around the assembly. When the assembly is baked a mould of great precision and fine surface finish is formed in a solid block of ceramic, with the wax melting out as before.

The metal is poured and allowed to solidify, and the mould is then broken up to free the castings, which are accurate in size and have good surface finish. The method lends itself to the production of castings which would normally present difficulties through the high melting-point of the metal used. Fig. 2.10 shows such a component cast in stainless steel. The limits on the casting are ± 0.10 mm and, if produced from solid bar stock, the finished component would weigh only 20% of the original bar. To machine away so much material would be uneconomic and, in stainless steel, would be extremely difficult.

SUMMARY

Precision castings are required in a variety of materials ranging from the low melting-point zinc-based alloys to the sophisticated Nimonic alloys and stainless steels. The low melting-point materials are produced in permanent metal moulds and dies, and can be cast in gravity dies or under pressure. Pressure die-castings in zinc-based alloys are produced in hot-chamber machines which have the melting equipment incorporated in the machine itself. Aluminium and magnesium alloys, if cast in this type of equipment, tend to pick up iron from the moulds should the dies get too hot. They are therefore cast in cold-chamber machines where the hot metal is in contact with the equipment only while a shot is being made.

Die-castings are frequently used where the applied loads are small and appearance is important, e.g., decorative brightware.

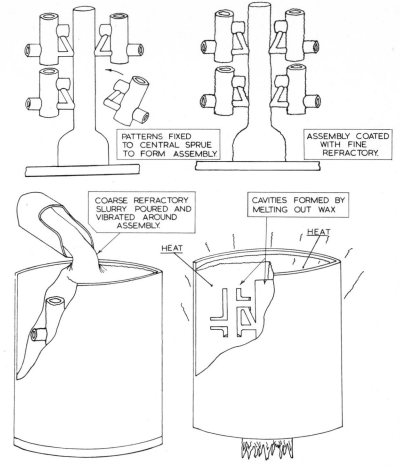

Fig. 2.9. Investment casting sprue assembly. (Reproduced by courtesy of British Investment Casting Manufacturers' Association.)

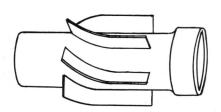

Fig. 2.10. Typical investment casting in stainless steel

Shell moulding uses expendable moulds which can be made very quickly by semi-skilled labour, and which produce accurate well-finished castings in a wide range of materials. The accuracy and finish are not quite as high as in die-casting and the process is slower but, compared with conventional sand-casting methods, output is very high.

Investment casting produces precise well-finished castings from expendable moulds in very high melting-point materials. The castings produced require very little machining and are

54

often of heat and creep-resistant alloys whose melting temperature is so high that a mould of refractory material is required.

A comparison of casting methods must consider such factors as cost of pattern, cost of mould, speed of process, dimensional accuracy, surface finish and any limitations. These factors are compared in the table below.

chamber process for those which have lower melting-points.

3. List the materials which may be used for die-casting purposes and give an example of a product made by each process.

A particular method is used for die-casting brass. Explain why this is necessary and describe the process.

Moulding method

Factor	Sand casting	Shell moulding	Die casting	Investment casting
Cost of pattern	Low	High	—	High
Cost of mould	High	Low	High	High
Speed of process	Slow	High	Very high	Slow
Surface finish and accuracy	Poor ± 0.40 mm up to 100 mm	Good ± 0.10 mm up to 100 mm	Very good ± 0.03 mm up to 100 mm	Very good ± 0.03 mm up to 100 mm
Comments	Medium and low production of ferrous and non-ferrous castings	High production of ferrous and non-ferrous castings	High production of castings in low melting-point alloys	Production of precision castings in high melting-point alloys

QUESTIONS

1. A circular container 450 mm diameter × 180 mm deep, having a wall thickness of 20 mm as shown in fig. 2.11, is to be cast in aluminium alloy. Batches of 800 are required at weekly intervals and at various positions around the top of the box are flanges, the position of which may be required to be changed for different batches. Suggest the best method of die-casting for such a component giving reasons for your choice.

2. Explain, with the aid of diagrams, the difference between the cold chamber and hot chamber die-casting processes. Give reasons for the cold chamber process being used for higher melting-point alloys, and the hot

4. The following components are all to be made by die-casting; select a suitable material and process giving reasons for your choice:
   (a) Cam shaft housing for a sports car engine.
   (b) Car door handle.
   (c) Sports car wheels.

5. List three advantages of die-casting over sand casting, other than the speed of production.

6. The box in question 1 is fitted with a cover, shown in fig. 2.11, held down by 12 bolts, which must seal against a gasket. The cover is to be removed regularly for servicing and it is felt that threads cut in the aluminium alloy box would wear or strip in use. Explain with the aid of a sketch how this problem could be overcome.

7. For another application, the cover plate shown in fig. 2.11 is to be cast in iron by the shell moulding process. List the stages in the manufacture of the casting, illustrating each stage with clear sketches of the equipment used.

8. A particular advantage of the investment casting process over all other methods of casting is in the complexity of the parts which may be produced by the process. Explain why this is so. List three other reasons for using the lost-wax process rather than die-casting or shell moulding.

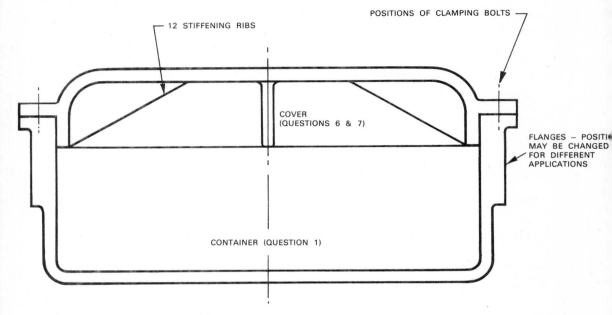

Fig. 2.11. Questions 1, 6 and 7.

3 Limits and fits and limit gauges

General Objective: *The student should understand a system of limits and fits for engineering.*

Consider fig. 3.1 – it represents a casting bored to 25.000 mm diameter, into which a bush of 25.100 mm diameter is to be pressed. The bush is to be machined to 16.000 mm inside diameter in which a shaft of 15.975 mm is to run. The designer has been very careful to design the parts to give the required strength, stiffness and other characteristics he considers necessary for the assembly to function correctly but, unfortunately, it is impossible to make. The drawing calls for exact sizes and although technological man can make and measure parts to an accuracy of $\pm 0.000\,05$ mm under certain ideal conditions he cannot make a part to an exact size except by chance and if he is lucky enough to hit the required size exactly he cannot measure it accurately enough to prove it.

The problem of how to make the parts to give the required assembly has three solutions:

1. Making to suit.
2. Selective assembly.
3. Using a system of limits and fits.

1. MAKING TO SUIT

One part is made as close to its designed size as possible with the equipment available, measured and then using a combination of machining, measuring and trying to fit the two together an approximation to the required fit is obtained. This method suffers the following limitations as far as economic high production is concerned:

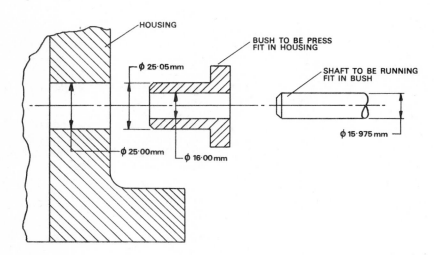

Fig. 3.1. Housing, bush and shaft with 'ideal' dimensions which cannot be made exactly as dimensioned.

(a) skilled craftsmen are required to obtain the desired fit;

(b) parts cannot be manufactured in bulk and selected at random for assembly;

(c) parts must be produced one at a time and production is slow and costly;

(d) quality of fits produced varies by unknown amounts;

(e) replacement spares cannot be made in advance and stored.

SELECTIVE ASSEMBLY

In this case the parts are all measured and graded into groups of closer limits than the full range of limits which the machines are producing. We may have five grades[1] of shaft and five grades of hole, the grades being numbered. If a grade 5 shaft is assembled into a grade 5 hole the assembly will function correctly. If a colour code method is used, a green-code hole is always assembled with a green-code shaft, and so on.

It is worth examining this system rather more deeply. Why, when a machine is operating at a fixed setting, does it produce parts of different size? The size of a given part selected at random depends to some extent on a great many uncontrollable minor factors, no single one of which will produce a measurable change in the size of the part. Usually, some factors tend to produce an oversize part and some an undersize part; their effects cancel out and most parts are therefore close to the set size. Occasionally, more of these factors tend to oversize than undersize and an oversize part is produced, and vice-versa.

If then we take a batch of shafts from a machine whose work has a random variation of 0.025 mm and measure them to the nearest 0.005 mm, we should have five size groups, and most of the parts will be in the centre group, as shown in fig. 3.2. If we repeat this procedure with a batch of holes we see the same pattern emerging.

[1] The term 'grade' here should not be confused with the IT grades in BS 4500.

If we now arrange for grade I shafts to be mated with grade I holes, grade II shafts with grade II holes and so on, it can be seen from fig. 3.2 that there will be about the same numbers of each, provided the machine settings are controlled carefully.

This system of selective assembly is generally used where two conditions prevail:

(a) The parts cannot be made economically to the required accuracy but can readily be measured and graded.

(b) The assembly is replaced as a complete unit when necessary, not repaired by replacing individual parts.

Typical applications of selective assembly are in ball and roller bearing manufacture and in cylinder bores and pistons, and pistons and gudgeon pins, in motor car engines.

Although selective assembly overcomes some of the problems of making to suit in that it allows economic production methods to be used with, in most cases, unskilled operators, the work involved in measuring, grading, storing in graded batches and selecting for assembly all add to the cost of the finished assembly.

3. USING A SYSTEM OF LIMITS AND FITS

Specific Objective: *The student states and gives reasons for adopting a system of limits and fits.*

Ideally a method of production is required in which:

(a) The most economic methods of manufacturing the parts can be used. Such methods usually involve the use of automatic or semi-automatic machines run by semi-skilled operators.

(b) All parts are completely interchangeable, i.e. if one bin contains 1000 shafts and another 1000 holes any pair of parts selected at random will go together to make a satisfactory assembly.

(c) All assemblies thus produced will be

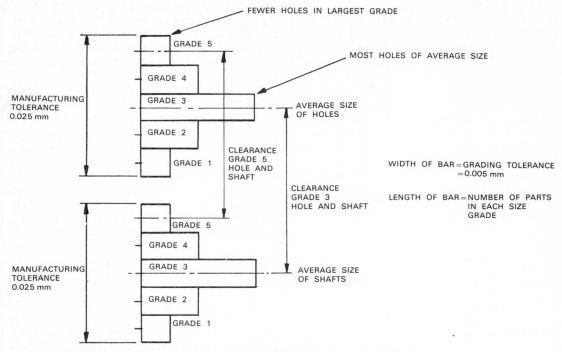

FEWER HOLES IN LARGEST GRADE

GRADE 5

MOST HOLES OF AVERAGE SIZE

GRADE 4

GRADE 3

MANUFACTURING TOLERANCE 0.025 mm

AVERAGE SIZE OF HOLES

GRADE 2

GRADE 1

CLEARANCE GRADE 5 HOLE AND SHAFT

WIDTH OF BAR = GRADING TOLERANCE = 0.005 mm

CLEARANCE GRADE 3 HOLE AND SHAFT

LENGTH OF BAR = NUMBER OF PARTS IN EACH SIZE GRADE

GRADE 5

GRADE 4

GRADE 3

MANUFACTURING TOLERANCE 0.025 mm

AVERAGE SIZE OF SHAFTS

GRADE 2

GRADE 1

Fig. 3.2. Relationships between grades of holes and shafts in selective assembly.

uniformly acceptable. There will be some variation in the fits produced but the amount of variation will have been predetermined by the designer so that the assembly will function correctly throughout its design life.

Compared with making to suit or selective assembly the advantages of such a system's cost and/or quality are enormous. It is not even necessary to measure the parts; limit gauges can be used to ensure that the parts are within the size limits fixed by the designer and thus inspection of the parts becomes, at most, a semi-skilled operation, further reducing the costs.

Consider again the parts in fig. 3.1 to see how the design may be modified to enable such a manufacturing method to be used. Having fixed the basic sizes the designer must decide what is the maximum variation in the size of each part which can be *tolerated*. By applying this *tolerance*, or variation in size to the basic size

the maximum and minimum *limits* of size can be arrived at and if these are chosen correctly all of the requirements specified will have been satisfied. The thinking in arriving at the limits will be as follows:

(a) *Housing bore*
Using a normal boring process we can achieve, let us say, a tolerance of 0.025 mm. Thus if the minimum diameter of the hole is to be 25.000 mm the maximum diameter will be 25.025 mm.

(b) *Outside diameter of bush*
We should be able to turn this to a tolerance of 0.025 mm. If the minimum interference to hold the bush in the bore is 0.050 mm and the maximum diameter of the hole is 25.025 mm it follows that the minimum diameter of the bush will be 25.075 mm. To this is added the tolerance of 0.025 mm to give the maximum O.D. for the bush of 25.100 mm.

59

(c) *Inside diameter of bush*
If this is reamed a tolerance of 0.025 mm should be achievable. If the minimum diameter is 16.000 mm the maximum diameter will be 16.025 mm.

(d) *Shaft diameter*
If we grind the shaft we should be able to hold a tolerance of 0.012 mm but the minimum clearance is to be 0.025 mm, so the maximum diameter of the shaft will be 15.975 mm and the minimum will be 15.963 mm.

These limited dimensions are shown in fig. 3.3. Note that by comparing the maximum and minimum conditions, as shown in the table below, we hold our original minimum values, but due to the tolerances, maximum values which far exceed these are to be allowed. This condition cannot be avoided and the designer must produce a compromise in which the worst conditions still allow correct functioning of the assembly.

TYPES OF FIT

Specific Objective: *The student should be able to define the three principle classifications of fit.*

The above example illustrates two types of fit, clearance and interference. There is a third, the transition fit, which can be clearance or interference in cases where a 'large' hole mates with a 'small' shaft or a 'small' hole mates with a

PART				
Dimension	Housing	Bush O.D.	Shaft O.D.	Bush O.D.
Maximum Dia.	25.025 mm	25.100 mm	15.975 mm	16.025 mm
Minimum Dia.	25.00 mm	25.075 mm	15.963 mm	16.000 mm
Tolerance	0.025 mm	0.025 mm	0.012 mm	0.025 mm
Limiting conditions	Max. interference 0.100 mm Min. interference 0.050 mm		Max. clearance 0.052 mm Min. clearance 0.025 mm	

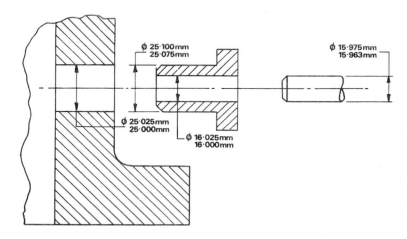

Fig. 3.3. Housing, bush and shaft with limits on dimensions. These parts can now be made.

'large' shaft. This fit is used for locations, dowels etc. so that parts can be easily assembled without undue pressure, and is sometimes known as a 'push' fit. The three types of fit can be clearly defined as follows, fig. 3.4 illustrating them diagrammatically.

he would hardly get any work done. Systems of limits and fits have therefore been developed in which the limits and tolerances for mating parts are set out in tabular form. The standard system in Britain is the British Standard 4500, an extract from which is shown in Appendix I.

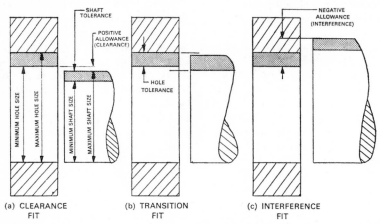

Fig. 3.4. Three basic types of fit, hole-base system. Note that the hole limits are the same in each case.

1. *Clearance Fit:* The largest shaft must be smaller than the smallest hole.
2. *Transition Fit:* The largest shaft is larger than the smallest hole and the smallest shaft is smaller than the largest hole (in fig. 3.4.(b) the tolerance zones overlap).
3. *Interference Fit:* The smallest shaft is larger than the largest hole.

Other definitions which are used are:
4. *Limits:* The maximum and minimum sizes of the part, which must not be exceeded.
5. *Tolerance:* The amount of variation in part size which can be tolerated, i.e., the difference between the limits.
6. *Allowance:* The prescribed amount between the mating pair in the maximum metal condition. A positive allowance gives clearance and a negative allowance gives interference.

SYSTEMS OF LIMITS AND FITS

If a designer had to perform the computation shown on pages 59–60 for every fit he required,

Here we shall attempt to analyse the thinking behind such systems and the student should refer to these tables to see how they support this reasoning.

Two distinct bases can be used in producing a system of limits and fits. These are:
1. Hole-basis system;
2. Shaft-basis system.

1. HOLE-BASIS SYSTEM

For a given size, the limits on the hole are always the same and a series of fits is obtained by applying different limits to the shafts. Thus, for a nominal dimension of 25.00 mm diameter, all holes will be, say, $25.00 {\,}^{+0.025}_{-0.000}$ mm diameter, and in conjunction with this common hole we obtain fits as follows:

(a) *Interference fit:* Shaft size $25 {\,}^{+0.075}_{+0.050}$ mm diameter, giving maximum and minimum interference of 0.075 mm and 0.025 mm respectively.

61

(b) *Transition fit:* Shaft size $25.00 {}^{+0.025}_{-0.000}$ mm, giving maximum clearance *and* maximum interference of 0.025 mm.

(c) *Clearance fit:* Shaft size of $25.00 {}^{-0.025}_{-0.050}$ mm, giving maximum and minimum clearances of 0.075 mm and 0.025 mm respectively.

2. SHAFT-BASIS SYSTEM

For a given size, the limits on the shaft are always the same and a series of fits is obtained by varying the size of the holes.

These two systems are shown diagrammatically in figs. 3.4 and 3.5. The hole basis is most commonly used since many holes are made with fixed-size tooling, i.e., drills, reamers etc. If a shaft-basis system with 12 different fits was used we might need 12 different reamers at each nominal size, which would be very expensive apart from problems of identification. Shafts, however, are made with adjustable tooling, i.e., lathes, grinding machines etc., and a range of different-sized shafts can readily be produced.

Types of hole

Even with a hole-basis system, it is found that just one class of hole is not practicable. If one class only was used it would have to be to the finest tolerance required, say for fine grinding and lapping, and be specified for all work. It would be ridiculous if we applied a tolerance of 0.001 mm to a hole that was to be flame-cut! Thus, even with a hole-basis system, a range of holes is required to permit economic methods of manufacture.

Examine the extract from B.S. 4500 in the Appendices and see how many types of hole are used.

Size of work

As the work size increases it is not possible, nor usually necessary to hold the same tolerances; larger limits are provided on larger work sizes.

Do the tolerances increase with work size in B.S. 4500?

VARIATION IN HOLE LIMITS

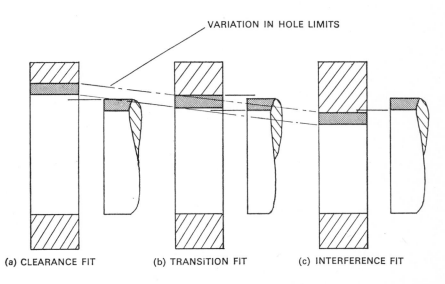

(a) CLEARANCE FIT   (b) TRANSITION FIT   (c) INTERFERENCE FIT

Fig. 3.5. Shaft basis system of obtaining different fits. Note that the shaft limits are the same in each case.

B.S. 4500 (1969): LIMITS AND FITS FOR ENGINEERING

Specific Objective: *The student should be able to define the terminology in common use in B.S. 4500, and relate manufacturing processes to tolerance grades.*

This is a comprehensive system to cover all classes of work from fine gauge making to heavy engineering. It takes into account size of work and class of work, and provides for hole-basis systems or shaft-basis systems as required. It must be emphasised that no organisation is expected to use the complete system, but to extract a sub-system for its own use, and we shall see how this can be done.

Size of work
The tolerances are set out in size steps in tabular form for all classes of work. For a given class of work and type of fit, one looks for a size step containing the nominal size required. Thus, our shaft and bush nominal size of 16.00 mm diameter on p. 60 falls in the size step 10 mm to 18 mm, and the bush and housing nominal diameter of 25 mm is in the size step 18 mm to 30 mm.

Class of work
Eighteen grades of tolerance are quoted, designated from 00, 0, 1 to 16 by numbers, and suit the types of work shown below.

Type of fit
The position of the tolerance zone relative to the nominal size is indicated by a letter, capital letters being used for holes and small letters for shafts. Twenty-eight holes and 28 shafts are specified, their positions relative to the nominal size being shown on page 64.

Class **H** is most important since one of its limits is the nominal size and the other above it. It follows that, when used in conjunction with a class **H** hole, shafts of class **a** to **g** give clearance fits, classes **h, js, j** and **k** give transition fits and classes **m** to **zc** give interference fits.

Each of these classifications may be associated with a given grade of tolerance and a hole may be designated **H7, H11, H14** etc. All these holes have a low limit of + 0.0000, the

Grade of Tolerance	Class of Work
00, 0, 1	Gauge blocks.
2	High quality gauges. Plug gauges.
3	Good quality gauges. Gap gauges.
4	Gauges. Precise fits produced by lapping.
5	Ball bearings. Machine lapping. Fine boring and grinding.
6	Grinding. Fine boring.
7	High quality turning. Broaching. Boring.
8	Centre-lathe turning and boring. Reaming. Capstan lathes in good condition.
9	Worn capstan or automatic lathes. Boring machines.
10	Milling, slotting, planing, rolling, extrusion.
11	Drilling, rough turning and boring. Precision tube drawing.
12	Light press work. Tube drawing.
13	Press work. Tube rolling.
14	Die casting or moulding. Rubber moulding.
15	Stamping.
16	Sand casting. Flame cutting.

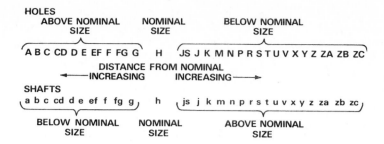

high limit increasing as the number increases. Similarly, shafts may be designated **a10**, **e8**, **p9** etc. If associated with a class **H** hole, a shaft **a10** would give a clearance fit, **e8** a closer clearance fit, and **p9** an interference fit. A typical fit is therefore designated **H7/p9** or **H7 − p9**.

Practical application of B.S. 4500

Specific Objective: *The student should be able to select values from B.S. 4500 and determine limiting values.*

Let us consider a light machine shop doing grinding, turning, milling and shaping, i.e., general machining, for which a system is to be extracted from B.S. 4500. It is decided to use a *hole-basis system* and this means that all holes will be of class **H**. The grades of tolerance used, according to the class of work, will be, say **H7** and **H8**, which should cover most cases. The work will require press fits using class **p** shafts, a transition fit using class **k** and three clearance

fits, **e** for loose clearance, **f** for medium running and **g** for precise location.

Analysing this further, close limits would not be used for class **e**, so all **e** shafts will be of grade **8** and all loose clearance fits will be **H8/e8**.

Medium-running fits will use an **H9** hole in conjunction with an **f8** shaft and precise location **H7/g6**.

A similar type of analysis is made for transition and interference fits, and a complete system suitable for the class of work can be reproduced in a similar form to that of Appendix I, which is a data sheet extracted from B.S. 4500 to cover general engineering work.

Using B.S. 4500 let us now dimension our housing, bush and shaft assembly from p. 59 and calculate tolerances and maximum and minimum clearances, the results being shown below in tabular form. Those for the interference fit would require machining to much tighter tolerances than the previous example on page 59.

	H7	p6	H8	f7
Dimension	Housing bore	Bush O.D.	Bush I.D.	Shaft O.D.
Maximum dia.	25.021	25.035	16.027	15.984
Minimum dia.	25.000	25.022	16.000	15.950
Tolerance	0.021	0.013	0.027	0.034
Limiting conditions	Maximum interference 0.035 Minimum interference 0.001		Max. clearance 0.077 Min. clearance 0.016	

Finally, the terminology of **H7/p6** etc. is only intended for use by the designer. For workshop drawings, the draughtsman then usually converts such specifications into dimensions, from the data sheet. A drawing which showed a hole diameter as 30.00 mm **H11**, sent to the workshop, might result in a swift trip to the drawing office with a demand, in no uncertain terms, for more information.

LIMIT GAUGES

Specific Objective: *The student should be able to apply the principles of gauge design as set out in the appropriate British Standards.*

Adoption of a system of limits and fits logically leads to the use of limit gauges, with which no attempt is made to determine the size of a workpiece – they are simply used to find whether or not the component is within the specified limits of size. The simplest forms of limit gauges are *plug gauges*, used for checking holes, and *gap gauges* and *ring gauges*, both of which are used for checking the diameter of shafts.

Consider first a hole on which the limits of diameter are specified. It would appear that the plug gauge would consist of two cylinders, the GO gauge having a diameter equal to the minimum hole size or *maximum metal limit* and the NOT GO gauge having a diameter equal to the maximum hole size or *minimum metal limit*.[1] If the GO gauge enters the hole then the size of the hole is above the maximum metal limit. If the NOT GO gauge will not enter the hole then its size is less than the minimum metal limit. It follows that the hole size is within limits. Note that the GO gauge always checks the maximum metal limit and the NOT GO gauge always checks the minimum metal limit.

[1] The terms *maximum and minimum metal limit* are used frequently in work on limits and fits and limit gauges. They enable certain rules of gauge design to be universally applied to the gauging of both holes and shafts.

Unfortunately it is not as simple as this: nothing can be made to an exact size or measured with absolute precision and this includes gauges. Thus the gauge maker requires a tolerance to which to work and the positioning of this tolerance relative to the basic size of the gauge fixes the gauge limits of size. For instance, if the gauge tolerance allows the size of a plain plug GO gauge to be increased and that of a NOT GO gauge to be decreased, the gauge will tend to reject work which is just within the upper and lower limits. Similarly, if the gauge tolerance allows for an increase in size of a NOT GO plug gauge and a decrease in size of the GO gauge then the gauge will tend to accept work which is just outside the limits.

It follows that a number of questions must be answered in designing a simple limit gauge.

(a) What magnitude of tolerance shall be applied to the gauge?

(b) How shall the tolerance zones for the gauge be disposed relative to the tolerance zones for the work?

(c) What allowance shall be made for the gauge to wear?

These considerations are all dealt with in B.S. 4500: Part 2: 1974: *Inspection of Plain Workpieces*.[1]

(a) GAUGE TOLERANCES

As a general guide in any measuring or gauging situation the accuracy of a piece of measuring equipment needs to be ten times as good as the tolerance on the work it is designed to measure. This means that the gauge tolerance will be approximately 10% of the tolerance on the work it is designed to gauge, the gauge tolerance being rounded to the nearest 0.001 mm unit.

Examination of the tolerance grades discussed on p. 63 shows that they are arranged in what is known as a 'preferred number series', this particular series being

[1] This standard largely supersedes the old standard, B.S. 969, but that standard will remain in being for an interim period to cater for gauges still in use and designed to it.

65

grouped in sets of five so that the magnitude of any tolerance grade is ten times that of its equivalent in the previous group. Thus, for size step 30 − 50 mm the tolerance grades are as follows:

Grade 1 = 0.0015 mm	Grade 6 = 0.016 mm	Grade 11 = 0.160 mm
2 = 0.0025 mm	7 = 0.025 mm	12 = 0.250 mm
3 = 0.004 mm	8 = 0.039 mm	13 = 0.390 mm
4 = 0.007 mm	9 = 0.062 mm	14 = 0.620 mm
5 = 0.011 mm	10 = 0.100 mm	15 = 1.000 mm

In this table grades 01, 0 and 16 have been omitted, but it can be seen that if a plug gauge is required for a hole specified as ϕ 25 H11 it is seen that the gauge tolerance will be to grade 6. This is not strictly followed in B.S. 4500: Part 2, but the principle of relating gauge tolerances to work tolerances is shown in table 1 of that standard. It is interesting to note that the table not only specifies the tolerance on the size of the gauge but the allowable errors in gauge form, and that the tolerance magnitude varies with the type of gauge according to the difficulty of manufacture.

(b) DISPOSITION OF GAUGE TOLERANCES

Having determined the tolerance magnitude for the gauge it must now be positioned relative to the work limits, so that it does not tend to accept defective work but at the same time does not tend to deprive the production department excessively of its work tolerance. In B.S. 4500: part 2 the disposition of the gauge tolerances is specified as follows:

The tolerance on the GO gauge is disposed bilaterally (equally on either side) about a line set a distance z (or z_1 for ring gauges) from the maximum metal limit of the work and within the tolerance zone of the work.

The tolerance on the NOT GO gauge is bilaterally disposed about the minimum metal limit of the work.

(c) WEAR ALLOWANCE

An allowance for wear is normally only applied to GO gauges. A NOT GO gauge should rarely be fully engaged with the work and should therefore suffer little wear. The allowance for wear on new GO gauges is made by placing the tolerance zone for the gauge within the tolerance zone of the work. It then remains to decide how much wear may be allowed to take place before the gauge should be considered unfit for further use. This is specified by a line positioned y (or y_1) from the maximum metal limit and outside the tolerance zone for the work.

These tolerance dispositions and allowances are shown in figs. 3.6 and 3.7. The design sizes for a limit gauge can be determined by the following method which, for the purposes of illustration, requires GO/NOT GO gap gauges for a shaft specified as ϕ45–h10.

1. Find the limits of size for the shaft. From table 1 in B.S. 4500 the tolerance is 0.100 mm. As grade 'h' shafts have limits of $^{+0}_{-x}$ the limits will be 45.000/44.900 mm.

2. Refer to table 2 in B.S. 4500: part 2, to find the following

Gauge tolerance grade 4 = 0.007 mm
$z_1 = 0.011$ mm
$y_1 = 0.000$ mm

3. Combine this information with that obtained for the work

GO gauge limits $= (45.000 − z_1) \pm 0.0035$ mm
$= 44.989 \pm 0.0035$ mm
NOT GO gauge limits $= 44.900 \pm 0.0035$ mm

For the GO gauge the maximum size to which it may be allowed to wear is the maximum metal limit of the work, i.e. 45.000 mm.

Note that we have shown earlier that the gauge tolerance should be 10% of the work tolerance and the tolerance grade for the gauge should therefore be grade 5. However where the work tolerance is rather large, as in this case, the tolerance grade for the gauge can go down one grade. The assumption of 10% should only be used as a guide if B.S. 4500: part 2 is not available.

For gauges up to and including 180 mm in size this procedure is simplified in tables 3 and 4 of B.S. 4500: Part 2 which gives the gauge limits and wear limit direct, relative to the maximum and minimum metal limits of the work. These must be determined from B.S. 4500: Part 1.

CONSTRUCTION OF GAUGES

Specific Objective: *The student should be able to sketch and describe the features and setting of plain and adjustable gauges.*

B.S. 4500: Part 2 sets out only the limits and tolerances for limit gauges. The actual design of simple gauges is set out in B.S. 1044: *Recommended Designs for Plug, Ring and Gap Gauges.*

PLUG GAUGES

The gauging members of plug gauges may be of plain carbon steel, hardened and stabilised, or of good quality case-hardening mild steel. There is little to choose between the materials and the one selected may depend upon

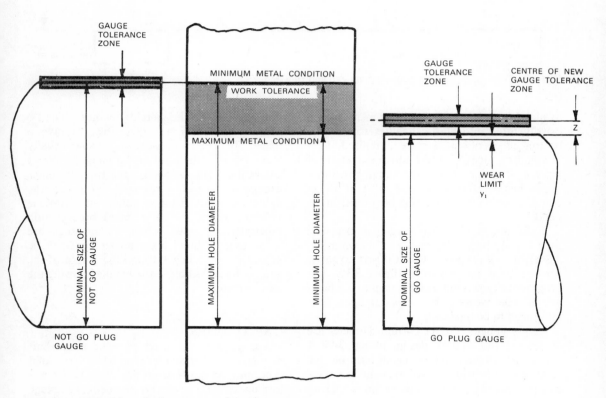

Fig. 3.6. Disposition of tolerances on plain plug gauges.

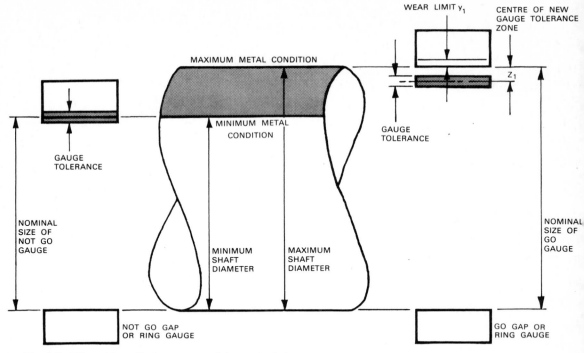

Fig. 3.7. Disposition of tolerances on plain gap and ring gauges.

availability and cost. The gauging members are usually fitted into plastic handles so that the heating effects of handling are minimised.

Smaller gauges, up to 64 mm diameter are of *taper-lock* design, that is, the gauging member shank has a self-locking taper of 1 in 48 on diameter which fits into a matching taper in the handle.

Separate GO and NOT GO gauging units can be used in a double-ended gauge as shown in fig. 3.8(a), or a *single* progressive GO/NOT GO gauge can be used as shown in fig. 3.8(b). The progressive type is more convenient to use but if one member wears or becomes damaged, both units need to be replaced.

Larger gauges of over 64 mm diameter are usually of the tri-lock design in fig. 3.9. A hexagonal plastic handle is again used but the gauging unit is held in place by a special screw and prevented from rotating by three locking prongs on the end of the handle which engage with grooves on the gauging member. The GO gauging members are reversible, to give a longer life. The initial wear on a gauge usually takes place at the nose of the gauge, where it enters the work, and when the wear becomes excessive the unit is reversed. Note that the lightening holes in larger gauges also provide venting for air which in blind holes would otherwise be trapped. Where smaller gauges are to be used in blind holes an air vent can be provided by drilling through the length of the gauge, the cross hole in the handle allowing the air to escape.

GAP GAUGES

These gauges are used for checking shaft diameters and may be adjustable. The solid type may be made from flat plate or from a forging: either type can be obtained from gauge manufacturers, finished and ready for use, or in

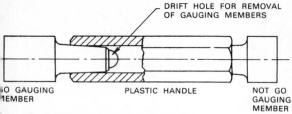

O GAUGING
MEMBER   PLASTIC HANDLE   NOT GO
GAUGING
MEMBER

Fig. 3.8. (a) Double-ended plain plug gauge.

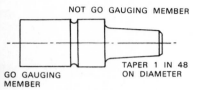

NOT GO GAUGING MEMBER

GO GAUGING
MEMBER   TAPER 1 IN 48
ON DIAMETER

Fig. 3.8. (b) Progressive plain plug gauge.

the soft state to be hardened, stabilised and finished to size by the user.

Adjustable gap gauges have a forged frame in which adjustable anvils are fitted and can be set to reference discs. With care, the anvils can be set to within ± 0.002 mm and thus allow full use to be made of the work tolerance, some of which may be taken up by the gauge tolerance in a solid gap gauge. The adjustment also allows

wear to be taken up, while if the anvil faces become non-flat they can be reground and re-adjusted. The anvils normally have a screw adjustment but do not themselves rotate with the screw. A separate device is used to lock the anvil in the frame and, after locking, the gauge is again checked. Finally the adjustment is sealed by a lead or wax plug as shown in fig. 3.10. Note the plastic grip fitted to reduce heat transfer from the hand to the gauge frame.

One limitation of gap gauges is that they will not detect a particular form of non-roundness called *lobing*. The simplest lobed figure has three lobes and is based on an equilateral triangle, as shown in fig. 3.11. From each corner of the triangle is struck a large radius 'R' and a small radius 'r' so that the large radius blends with the small radius at the other two corners. This figure will always enter a pair of gauging anvils set at $(R + r)$ apart, but the smallest hole it will enter is much larger than $(R + r)$. This condition will not be detected by any measurement involving the use of parallel anvils such as those of a micrometer or comparator. It can be detected by rotating the lobed work under a dial gauge or comparator when the work is resting on a vee block as shown in fig 3.12.

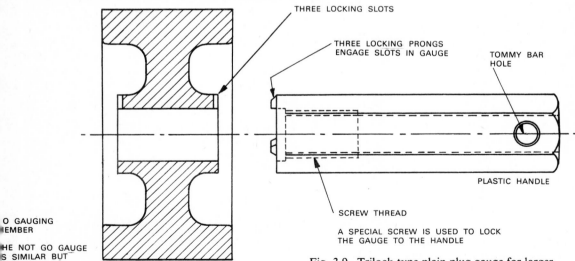

THREE LOCKING SLOTS

THREE LOCKING PRONGS
ENGAGE SLOTS IN GAUGE

TOMMY BAR
HOLE

PLASTIC HANDLE

O GAUGING
MEMBER

HE NOT GO GAUGE
S SIMILAR BUT
NARROWER

SCREW THREAD

A SPECIAL SCREW IS USED TO LOCK
THE GAUGE TO THE HANDLE

Fig. 3.9. Trilock-type plain plug gauge for larger rings.

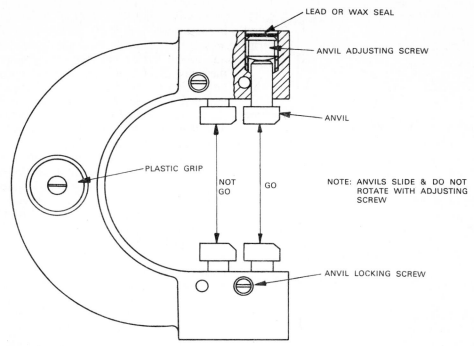

Fig. 3.10. Adjustable gap gauge.

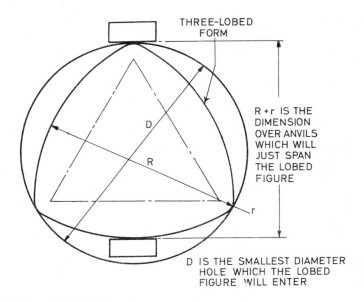

Fig. 3.11. Effect of lobing on cylindrical work.

70

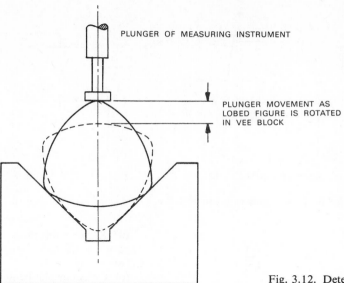

PLUNGER OF MEASURING INSTRUMENT

PLUNGER MOVEMENT AS LOBED FIGURE IS ROTATED IN VEE BLOCK

Fig. 3.12. Detection of lobing.

RING GAUGES

It can be seen from fig. 3.11 that a lobed figure will not enter a hole whose diameter is only the same as the gap between the anvil faces. If the machining process being used is likely to produce this form, e.g., with a capstan roller box or a centreless grinder incorrectly set, a GO ring gauge should be used as a periodic check. B.S. 1044 gives details of the overall dimensions of ring gauges required to ensure sufficient rigidity. The outside diameter is normally finished to a fine or medium knurl with a generous chamfer on the corners.

GAUGE MARKINGS

All gauges should be clearly marked with the size of the individual gauging members, the words GO or NOT GO, the type of gauge if necessary, e.g., General or Reference, and the manufacturer's name. For plug gauges with plastic handles, this information is engraved upon the handle and is more clearly visible.

THE GAUGING OF TAPERS

Tapers can be checked by taper limit gauges. This seems an obvious statement but it must be emphasised that such gauges *do not* check the angle of taper. They are designed only to check the diameter at a particular position along the taper, usually at one end.

These gauges are usually *step*, or *thumbnail* gauges, the limits of diameter being defined at the top and bottom respectively of a step which is ground at the appropriate end of the gauge. Fig. 3.13 shows how a taper ring gauge is used to check the diameter of an external taper, while fig. 3.14 shows the geometric details of a taper plug gauge. The semi-angle of taper, $\theta/2$, can be determined by using sine bar centres while the diameter D of the small end is obtained by measurement over rollers.

From fig. 3.15,

$$\tan \frac{\theta}{2} = \frac{D \max - D}{2H}$$

$$D \max = \left[2H \times \tan \frac{\theta}{2}\right] + D$$

and $D \min = \left[2(H-S) \times \tan \left(\frac{\theta}{2}\right)\right] + D$

where $S =$ step height.

71

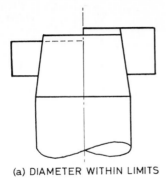

(a) DIAMETER WITHIN LIMITS

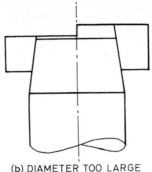

(b) DIAMETER TOO LARGE

Fig. 3.13. Method of using a taper ring gauge.

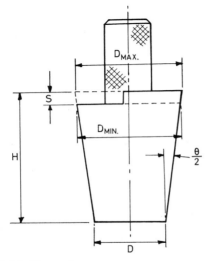

Fig. 3.14. Taper plug gauge.

It should be noted that a slight error in the angle of taper and/or the diameter makes a significant difference in the values of D max and D min, the limiting diameters, while an error in the length H has little effect, due to the angle of taper. For this reason it is suggested that the following method is adopted in making such gauges.

(a) Make the gauge overlength and to an angle as close as possible to that specified for the work.

(b) Remove the gauge from the machine and accurately measure the small diameter 'D' and the semi-angle $\theta/2$.

(c) Calculate the height 'H' and the step 'S' to give the required dimensions 'D max' and 'D min'.

(d) Grind the length to the calculated height 'H' and step 'S'.

HOLE DEPTH GAUGES

These are also often made as thumbnail gauges as shown in fig. 3.15. The gauge is cylindrical

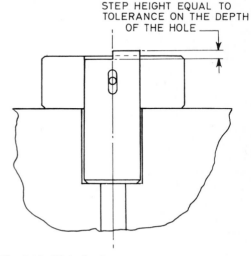

STEP HEIGHT EQUAL TO TOLERANCE ON THE DEPTH OF THE HOLE

Fig. 3.15. Hole depth gauge.

72

and is made a close sliding fit in the sleeve which is used to check the step and also enables the gauge to be removed from the hole. The gauge is retained in the sleeve by a pin fitting in an elongated slot, which allows the gauge to be pressed to the botttom of the hole.

So far we have dealt with the design of simple gauges in that their geometric form has been assumed, a plug gauge for a hole and gap or ring gauges for a shaft. However the shape or geometric form of limit gauges is not the simple matter it appears to be, even for these simple gauges and still less for limit gauges to be used for inspecting more complex components.

TAYLOR'S THEORY OF GAUGING

Specific Objective: *The student should be able to state Taylor's theory of gauging and explain examples of its use.*

Taylor's principle is the key to the geometric design of gauges and, if used correctly, enables the gauges to be designed so that they efficiently perform the function for which they are intended, that is, to identify work which is made according to the specification, and reject that which is outside the specification. Taylor's principle states that:

The GO gauge checks the maximum metal condition and should check as many dimensions as possible.
The NOT GO gauge checks the minimum metal condition and should only check one dimension.

It follows that, where possible, the GO gauge should be of *full-form*, the single gauge checking all GO dimensions. Separate NOT GO gauges should be used for each individual dimension.

Consider a system of limit gauges for the rectangular hole shown in fig. 3.16.

The GO gauge is used to ensure that the maximum metal condition is not exceeded and that metal does not encroach into the minimum allowable hole space. This can only be done by

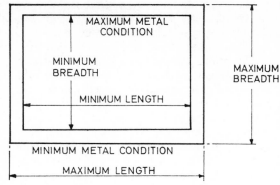

Fig. 3.16. Tolerance zone on a rectangular hole.

a gauge of full form made to the maximum metal condition, due allowance being made for gauge tolerance and wear as has been explained previously.

Now consider the NOT GO gauge. If this was made to gauge both length *and* breadth at the minimum metal condition (maximum hole size) a condition could arise where the breadth of the hole is within the specified limits and the gauge would not enter the hole, indicating that the component should be accepted. However, the length of the hole could be well outside the limits as shown in fig. 3.17 and the part should have been rejected.

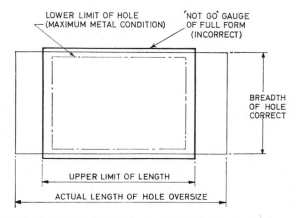

Fig. 3.17. Rectangular hole oversize in one direction. A full-form 'NOT GO' gauge will not reject such a hole.

73

Had separate NOT GO gauges been used for the two dimensions the fact that the length was outside the specified limits would have been detected and the part, quite rightly, rejected.

This principle should be applied to the design of all limit gauges if defective parts are not to be accepted by incorrectly designed gauges. We have seen (p. 70) how a lobed part will be accepted by a plain GO gap gauge. Taylor's principle explains why this is so. A shaft does not only have one diameter, it has an infinite number of diameters which should all be equal. A gap gauge only checks one diameter at a time and is therefore incorrect for use as a GO gauge. A GO ring gauge checks all the diameters at the same time and the error will be detected. Note that a plain gap gauge is perfectly correct for the NOT GO dimension.

In practice a GO/NOT GO gap gauge is used for convenience, but it can lead to problems and a GO ring gauge should be available in case of dispute. It is important to note that a NOT GO ring gauge is entirely unacceptable and should never be used.

A more complex situation arises in the gauging of screw threads. Consider a thread being cut by an adjustable die which has a pitch error, the thread being gauged using, *incorrectly* GO and NOT GO screw ring gauges both of which are of full form and full length of thread. When the thread diameter is judged to be about correct, it is offered up to the GO gauge and will not fully enter because of the pitch error. The die is adjusted to reduce the diameter until the thread enters the GO gauge and then offered up to the NOT GO gauge which it does not enter and the work is accepted. The reason the thread does not enter the NOT GO gauge is not because the diameter is correct but because the pitch is wrong. This is an outstanding example of incorrect work being accepted because the gauges have not been designed according to Taylor's principle. The NOT GO gauge is trying to assess major, minor and effective diameters as well as pitch and form. If the gauge is re-designed to check only the effective diameter, the thread would enter the NOT GO gauge and would rightly be rejected. This can be done by clearing the thread form on the gauge at crest and root and reducing the number of threads to reduce the effect of pitch errors. Thus the anvils of a NOT GO screw gap gauge are made as shown in fig. 3.18, the ideal system for limit gauging external threads being:

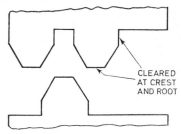

CLEARED AT CREST AND ROOT

Fig. 3.18. Forms of anvils on a 'NOT GO' screw gap gauge.

1. GO ring gauge of full form and full length of thread.
2. NOT GO gap gauge with anvils of the type shown in fig. 3.18.
3. Plain NOT GO gap gauge for the major (outside) diameter.

In practice a full form GO gap gauge is often used in conjunction with a NOT GO gap gauge of the type suggested for convenience and speed of operation. The work can be progressively passed through a GO gap gauge and offered directly up to the NOT GO gap gauge behind it. The only way to check work with a screw ring gauge is to screw it in and then unscrew it which all takes time. However, if both GO and NOT GO screw gap gauges are used they should be backed up with a GO screw ring gauge for periodic checks and particularly checking when the machine is being set up.

Using similar arguments it can be shown that an ideal gauging system for internal threads will be:

1. GO screw plug gauge of full form and full length of thread.
2. NOT GO screw plug gauge cleared at crest and root and having only two or three threads to check effective diameter only.
3. NOT GO plain plug gauge to check the minor diameter of the thread.

The major diameter is not gauged with a NOT GO gauge; it would be almost impossible and in practice the plain plug gauge is not usually used to allow the use of a double-ended screw-plug consisting of units 1. and 2. only.

Thus Taylor's principle is applied to gauge design for both simple and complex limit gauges. When a system of limit gauges is being designed the question should always be asked, 'Does it conform to Taylor's Principle?' If it does not it will almost certainly cause trouble.

MEASUREMENT AND GAUGING

Specific Objective: *The student should be able to compare the use of gauges with direct measurement.*

The use of systems of limits and fits coupled with the use of limit gauges were the two main factors that made possible interchangeable manufacture, and hence mass production as we know it. However, it must be realised that limit gauging has its limitations. When a piece of work is passed by a limit gauge all that is known is that the work is within the specified limits of size and little is known about the process being used to make the part. If the work is measured then the actual size of the workpiece, to within the limits of accuracy of the measuring equipment being used, is known. The size of the workpiece is an indication of the machine setting and if it is close to the work limit the operator knows that the machines will soon have to be reset or he will be producing scrap. Thus measurement gives considerably more information than gauging, not necessarily about the part, but about the process producing that part.

In modern quality control this idea has been considerably extended. It is accepted that the measured size of one part may not be representative of the process, so at regular intervals, a sample of the last five parts is taken off the machine and measured. The average size of the sample is an indication of the machine setting and by plotting these on chart, called a control chart, trends in changes in the machine setting can be observed and the machine reset before scrap is produced.

The reason that one part may not be representative of the whole is that no machine makes all parts the same even at the same setting. The amount of variation is called the *process variability* and is an indication of the precision of the machine. As the machine wears or something goes wrong the process variability gets worse and it may no longer be able to work to the limits required. The difference in size between the largest and smallest parts in the sample, called the *sample range*, is an indication of the process variability, and by plotting the sample range on a control chart the machine condition can be monitored, trouble being detected and corrected before it gets out of hand.

This technique is the basis of one form of *statistical quality control* and by its use, the production of scrap can be prevented (rather than defective parts being made and sorted out using gauges later). The use of these methods requires more highly skilled inspectors and may tie up some expensive measuring equipment so the decision to use measurement or limit gauging in a particular situation is largely a matter of economics. This, however, does not alter the fact that the amount of information gained by measurement is much greater than that obtained by gauging.

QUESTIONS

1. Give three reasons for using a system of limits and fits in modern industry compared with 'making to suit' or selective assembly. Under what conditions would selective assembly be used?

Give an example of its use.

2. With the aid of a block diagram define three principal classes of fit and state one use for each.

3. A fit is designated ϕ 35 H9/d10. Using the extract from B.S. 4500 provided in appendix I give:

    (i) shaft limits (34.920/34.820 mm);

    (ii) hole limits (35.062/35.000 mm);

    (iii) extremes of fit (Max. clearance = 0.242 mm; Min. clearance = 0.080 mm)

4. State the limits and manufacturing methods which would be used for features of workpieces specified as follows:

    (a) Shaft ϕ 52 f7, work not to be hardened.

    (b) Hole ϕ 18 H6 in hardened workpiece.

    (c) Hole ϕ 12-H11 in brass.

5. A hole is specified as ϕ 27.5 H11.
Sketch a limit gauge suitable for checking the hole and dimension the gauging given that:

$$z = 19 \mu m$$
$$y = 0 \mu m$$
Gauge tolerance = 10 μm

6. An octagonal hole is specified as 50 mm across flats to limits H9. Make dimensional sketches of suitable limit gauges for the hole, showing the basic gauge dimensions only.

    If $z = 11 \mu m$
$$y = 0 \mu m$$
gauge tolerance = 4 μm,
state the limits of size and wear for both the GO and NOT GO gauge.

7. A taper hole has an included angle of 15° and is checked by a 'thumb nail'-type taper limit gauge. Sketch the gauge and calculate the depth of the step to be ground on the large end if the tolerance on the large end diameter of the taper is 0.2 mm (0.76 mm).

4 Engineering measurements

The technician studying this book should already be familiar with the basic forms of measurement encountered in the workshop. In this volume we shall be considering some more sophisticated methods of measurement which have developed with the growth of modern technology.

Whatever method of measurement is used, there are certain principles which should be followed if errors are to be minimised. Even if every precaution is taken, no measurement is exact, and an estimate of the *accuracy of determination* of the measurement should always be made. For example, a gap gauge is measured by gauge blocks and is found to be 24.29 mm. This requires a combination of gauge blocks of 1.09 mm + 1.20 mm + 22.00 mm, each of whose individual errors may be negligible but in combination might be as much as ± 0.000 5 mm. The sensitivity of feel or touch must also be considered, plus the fact that gauge blocks are manufactured only in 0.01 mm increments and thus, by direct comparison, we cannot subdivide this unit. It follows that our measurement cannot be confidently stated to closer than 0.01 mm and should be presented as follows:

Gap size = 24.29 mm to an accuracy of determination of ± 0.01 mm

This may seem to be a longwinded way of putting it but at least everybody knows the true situation. The statement means, 'I have measured this gauge and find its size to be 24.29 mm but due to the possible existence of certain unavoidable errors it can be anything between 24.30 mm and 24.28 mm', which is longer still. (The gap size could be written '24.30 mm > Gap > 24.28 mm' but many people would not understand this.)

CAUSES OF ERROR IN FINE MEASUREMENT

1. TEMPERATURE
The standard temperature at which measurements should technically be made is 20°C (68°F) and the room in which fine measurements are carried out should be maintained as nearly as possible at this temperature. Although this is important for the finest measurements, it is in all cases much more important that all the components in a measuring system are at the same temperature. Thus if a measurement is made on a length bar comparator involving a comparison of two bars 400 mm long and all the components are at 18°C instead of 20°C, the error will not be significant. If, however, the measurement is made with one bar at 18°C and the other at 20°C the error incurred will be given by

$$\text{error} = l \, \alpha \, t$$
$$= 400 \times 0.000\,011 \times 2$$
$$= 0.008\,8 \text{ mm}$$

in which

l = nominal length = 400 mm
α = coefficient of linear expansion
= 0.000 011/°C
t = temperature difference = 2°C.

Thus the error due to different temperatures within the system is measurable and significant.

2. SINE AND COSINE ERRORS
These errors are due to misalignment of the measuring instrument and the work. A simple case is a measurement made with a dial gauge which is not parallel to the required line of measurement.
In fig. 4.1,

Dial gauge reading $= BC$
Actual measurement $= AB$

$$\frac{AB}{BC} = \text{Cos } \theta$$

where $\theta =$ angle of misalignment.

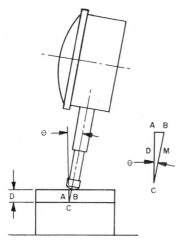

Fig. 4.1. Cosine error due to misalignment of measuring instrument.

3. PARALLAX ERRORS

These are due to the observer's line of sight not being normal to the dial of the instrument being read. In fig. 4.2, point A is the correct reading and point B is the incorrect reading due to the eye viewing at parallax angle θ.

$$\frac{AB}{AP} = \tan \theta$$

$$AB = AP \tan \theta$$

where AP $=$ distance from pointer to scale.

The parallax error AB will be reduced to zero if:

(a) The distance AP$=0$. This can be arranged by putting the pointer in the plane of the scale.

(b) The angle θ is zero. This is done by ensuring that viewing is normal to the scale.

Many instruments have a strip of mirror in the

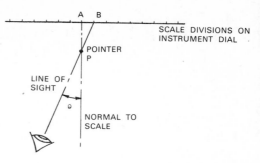

Fig. 4.2. Parallax error due to line of sight not being at right angles to scale of measuring instrument.

scale behind the pointer and the line of sight is at right angles to the scale when the pointer obscures its own image.

4. CALAMITOUS OR CATASTROPHIC ERRORS

These are not usually errors of measurement but errors of arithmetic. The inspector fails to carry 1 in adding a slip pile and an error of 1 mm is produced; a micrometer is misread by 1.00 mm or 0.50 mm; the radius of a roller is used instead of its diameter when measuring a taper. There is a simple method of avoiding these errors; if the component is measured by a rule as a rough check the measurement will be within the order of ± 0.25 mm. A greater difference than this between the final measurement and the rough check will generally be due to arithmetical error.

IMPROVED ACCURACY OF DETERMINATION

If a part is measured only once, the accuracy of determination is that of a single measurement. If a part is measured n times and the average measurement is then taken, the accuracy of determination of the average value is approximately $\pm x/\sqrt{n}$ where $\pm x$ is the accuracy of the single measurement. Thus if a measurement is made four times and the average of the four results is taken, the accuracy of determination of the average will be twice as good as that of the individual measurement.

It must be realised that this is only true if the complete measurement is repeated a number of times. It is not enough to set a comparator to gauge blocks once and then take a number of readings on the work alone. The setting must be displaced and the whole process repeated.

COMPARATIVE MEASUREMENT

General Objective: *The student should recognise the principle and use of various types of comparators.*

In the book[1] preceding this written for level 2 of this subject, the topic of comparative measurement was introduced and it was shown that it has certain advantages over direct measurement either by direct use of gauge blocks or by the direct reading of an instrument such as a micrometer. These advantages are:

1. The smallest increment measurable using gauge blocks to measure, say, a gap gauge directly is the smallest increment in the gauge blocks. That is, if the gauge blocks only have steps of 0.01 mm then that is the nearest value to which the measurement can be stated. A comparator enables these increments to be subdivided.
2. The measuring pressure can be better controlled using a comparator, the pressure being the same for all cases including the setting and reading from the instrument.
3. The range over which measurements are made and the chance of errors due to cumulative errors in the instrument movement is reduced.

In that previous work it was shown how the micrometer could be modified in various ways to be used as a comparator, how comparative measurements should be made and the results

[1] *Technician Manufacturing Technology 2* (Cassell, pp. 249–272).

set out and also showed the principle of operation of two types of mechanical comparator. For the sake of completeness, the work on mechanical comparators will be repeated here so that it can be easily compared with the sections on pneumatic and electrical comparators. The principles of optical comparators will be discussed in a later section (see p. 107) dealing with optical principles and methods of measurement.

In all comparative methods of measurement the instrument is first set to a standard, such as a pile of gauge blocks. A reading is then taken on the work, and the difference between the setting and the reading is the difference in size between the setting gauges and the workpiece. The size of the workpiece can thus be arrived at. This is explained in detail in the section on mechanical comparators and will not be repeated for the other types except where differences in the technique are necessary.

MECHANICAL COMPARATORS

Specific Objective: *The student should be able to explain, with the aid of sketches, the principles of operation of mechanical comparators.*

These are probably the commonest of the bench-type comparators and are extremely convenient in that they require no services such as electricity supplies or compressed air. They consist of a heavy base carrying the work stage and a column, up and down which the measuring head can be adjusted. Externally, the measuring head consists of the work contact plunger whose movement operates a pointer moving over a dial. Magnifications range from $500 \times$ to $5000 \times$, a typical instrument being shown in fig 4.3.

In use, a combination of gauge blocks is made up to the nominal size of the work to be measured. If the calibrated values of these blocks are known, the errors should also be totalled. Thus if the nominal size of the work is 51.57 mm we have:

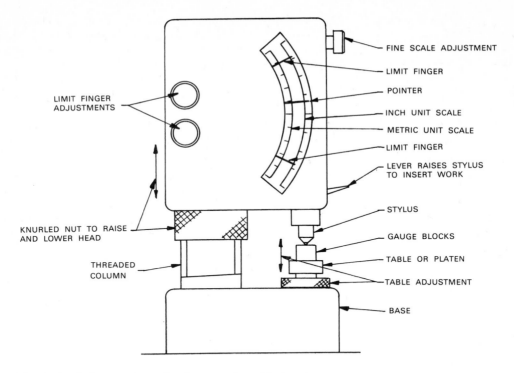

FINE SCALE ADJUSTMENT

LIMIT FINGER

POINTER

INCH UNIT SCALE

METRIC UNIT SCALE

LIMIT FINGER

LEVER RAISES STYLUS TO INSERT WORK

STYLUS

GAUGE BLOCKS

TABLE OR PLATEN

TABLE ADJUSTMENT

BASE

LIMIT FINGER ADJUSTMENTS

KNURLED NUT TO RAISE AND LOWER HEAD

THREADED COLUMN

Fig. 4.3. Mechanical comparator showing operating adjustments.

Gauge block (mm)	Known error (μm)
1.07	+ 0.03
1.50	+ 0.02
9.00	+ 0.05
40.00	+ 0.05
Total = 51.57 mm	+ 0.15 μm

If the comparator is calibrated with the smallest units of 0.000 1 mm it is now set at + 1.5 of these units, as shown in fig. 4.4. The errors in the gauge blocks are thus cancelled by presetting the comparator and when the workpiece is inserted its error is read directly.

The Sigma comparator is an example of a mechanical comparator whose amplification is obtained by means of a compound lever arrangement, a simplified form of which is shown in fig. 4.5. The overall magnification is the product of the two lever ratios, i.e.

$$\text{Magnification} = \frac{L}{x} \times \frac{R}{r}$$

The actual movement is shown in diagrammatic form in fig. 4.6. As the plunger is raised, the moving block follows the knife edge and the Y-shaped arm moves about the cross-strip pivot. A phosphor-bronze driving band is carried from the ends of the Y-arm about a driving drum of radius r to which is attached a pointer of length R. Thus as the Y-arm moves it causes the drum to rotate, carrying the pointer with it.

The Johansson comparator is a simple device based upon the same principle as that which operates the child's toy, in which a button is placed at the centre of a loop of twine, one end of which is hooked over each thumb. The button is now 'wound up' so that the twine is twisted in opposite directions on either side. If the thumbs are now alternately drawn apart and brought

80

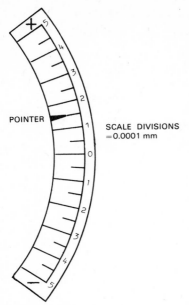

POINTER

SCALE DIVISIONS
=0.0001 mm

Fig. 4.4. High magnification comparator set to known error in gauge blocks.

together the button spins rapidly with a satisfying buzz. A little thought shows that a small linear movement of the thumbs produces a large angular movement of the button. Similarly, in a comparator, a small linear movement of the measuring stylus is required to produce a large angular movement of the pointer.

The Johansson Mikrokator shown in fig. 4.7 has the pointer mounted at the centre of a thin strip of metal permanently twisted in opposite directions on either side of the pointer. As the plunger is raised its movement is transmitted to the strip by the bell-crank lever arrangement, causing the strip to stretch and unwind, and the pointer to move across the scale.

PNEUMATIC COMPARATORS

Specific Objective: *The student should be able to explain with the aid of diagrams, the principles of operation and applications of pneumatic comparators.*

Pneumatic comparators or air gauges, as they are often called, work on the principle that if an

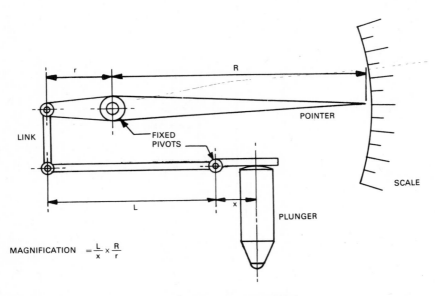

MAGNIFICATION $= \dfrac{L}{x} \times \dfrac{R}{r}$

Fig. 4.5. Essentials of compound lever type of comparator movement.

81

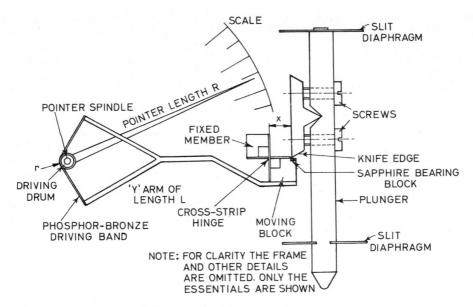

Fig. 4.6. Diagram of movement of Sigma mechanical comparator.

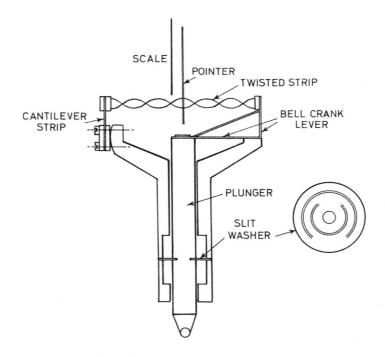

Fig. 4.7. Movement of Johansson Mikrokator.

air jet is brought into close proximity with a surface the flow of air is restricted, and can change both the pressure in the system and the rate of flow of air through the system. Both these changes are used to indicate the distance between the jet and the work face.

Back-pressure air gauges

Air gauges which indicate the change in a measured dimension by measuring a change in air pressure are called *back-pressure gauges*. They consist of two orifices, or jets, in series supplied with air at constant pressure, P_c. If the measuring jet is completely closed by the workpiece the air throughout the system remains at this pressure but if the measuring jet is gradually opened, by moving the workpiece, the back pressure between the jets, P_b falls. Thus by measuring the pressure between the jets we have a measure of the distance L in fig. 4.8. The pressure gauge is calibrated not in kN/m^2 but in the units of length in which L is to be measured.

Magnification of the measurement increases as the input pressure is increased and as the control orifice area is reduced.

A practical application of this instrument is shown in fig. 4.9. This is the type made by Solex Air Gauges Ltd, and shows an air plug gauge checking a bore.

The air from its normal source of supply, say the factory air line, is filtered and passes through a flow valve. Its pressure is then reduced and maintained at a constant value by a dip tube into a water chamber, the pressure value being determined by the head of the water displaced, with excess air escaping to atmosphere. The air at constant pressure then passes through the control orifice and escapes from the measuring orifice. The back pressure in the circuit is indicated by the head of water displaced in the manometer tube. The tube is graduated linearly to show changes of pressure resulting from changes in the measured dimension.

Another back-pressure air gauge is produced by Mercer Air Gauges Ltd, but this operates at the much higher pressure of 27.5 N/cm^2 gauge. The constant pressure input is produced from the line pressure by a diaphragm-type regulator and passed to the control orifice and thence to the measuring orifice.

Interesting features are:

(a) *Magnification adjustment.* This is achieved by means of a taper-needle valve in the

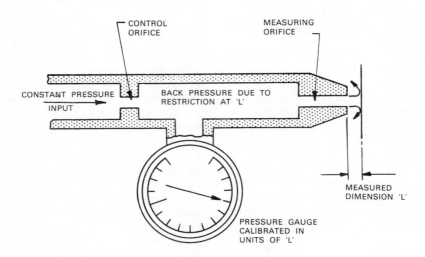

Fig. 4.8. Essentials of back-pressure type of air gauge.

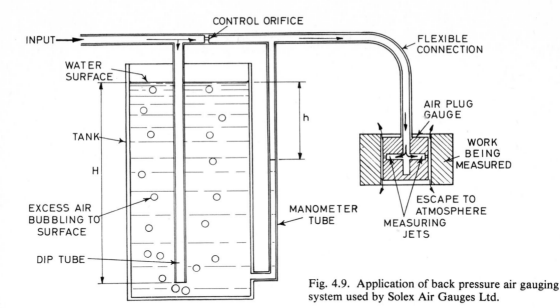

Fig. 4.9. Application of back pressure air gauging system used by Solex Air Gauges Ltd.

control orifice and enables a single scale to be used for all types of work by adjusting the magnification and zero settings.

(b) *Zero adjustment.* An air bleed, upstream of the measuring orifice and controlled by a taper needle valve, provides a zero adjustment.

The pressure measuring device is a Bourdon tube type of pressure gauge, the dial being graduated in linear units, i.e., 0.01 mm, 0.001 mm or inch units.

As with all other comparators, initial setting is by means of reference gauges.

Air gauges of this type (and electrical gauging units) have a distinct advantage over mechanical and optical comparators in that a small measuring unit can be remote from the amplifying and indicating head, and can also be conveniently built into a multiple gauging unit as shown in fig 4.10. The manometer tubes are set close together with coloured zones indicating the tolerance bands for the various dimensions and the operator can see at a glance whether all dimensions are within the limits or not.

A disadvantage of this type of comparator is that it does not have a linear response, i.e., a

graph of measured gap against reading is not a straight line. Great care must be taken to ensure that the scale is correct over its whole length and not just at the zero and extreme values.

Flow-velocity air gauges
In these air gauges, the measured gap between jet and workpiece is determined from the quantity of air passing through the system. Air enters the gauge at constant pressure into a tapered glass tube containing a 'float'. The float takes up that position in the tube where the upthrust of the air moving past it is equal to its own weight and will always take the same position at the same velocity of air flow. If the measured gap is large and a great quantity of air is flowing the float moves to the top of the tapered tube. If the gap is small, the quantity of air flowing is small and the float drops near to the bottom of the tube. The scale shown in fig. 4.11 is calibrated in 0.01 mm or 0.001 mm units, the magnification being altered by changing the tube to one of a different taper. Fine adjustments to the magnification are made by adjusting the supply pressure.

This type of air gauge lends itself to toolroom

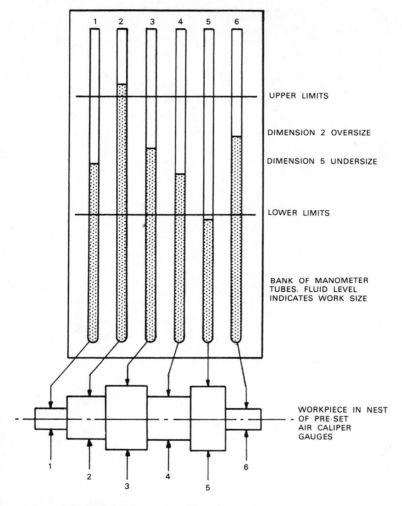

Fig. 4.10. Block diagram of multi-gauging set-up using air gauges.

inspection work, a height-setting micrometer, shown in fig. 4.12, being used as a setting master. The measuring jet is attached to a height gauge for convenience and differences in height between the work and the setting micrometer are shown on the air gauge. The height gauge used should preferably be of the type with a screw adjustment at the base of the column and no readings are taken from its scale; it is simply a convenient mount for the measuring jet unit.

Types of jet. The measuring jet may be a simple open jet, the air from which impinges directly upon the work. This has the advantage that it is self-cleansing, and the air will blow away small pieces of foreign matter and cutting fluid. Such jets are, however, prone to damage and the slightest burr on the jet edge will change

85

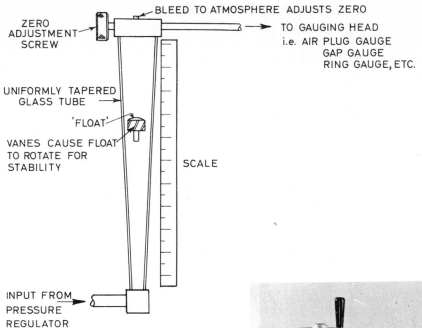

ZERO ADJUSTMENT SCREW

BLEED TO ATMOSPHERE ADJUSTS ZERO

TO GAUGING HEAD
i.e. AIR PLUG GAUGE
GAP GAUGE
RING GAUGE, ETC.

UNIFORMLY TAPERED GLASS TUBE

'FLOAT'

VANES CAUSE FLOAT TO ROTATE FOR STABILITY

SCALE

INPUT FROM PRESSURE REGULATOR

Fig. 4.11. Line diagram of flow-velocity-type pneumatic circuit.

the jet characteristics and cause incorrect readings. Open jets are normally used only where the nature of the gauge affords some degree of protection, as in the air plug gauge shown in fig. 4.9.

Note also that the reading from an open jet can vary with surface finish and geometry. The setting master and the work should, as far as possible, be of the same form and finish.

For measurements where an open jet might be damaged, the gauge incorporates a plunger to contact the work. Movement of the plunger causes a taper needle to move in an orifice and thus open or close the jet. Such a measuring head is shown in fig. 4.13 and may be built into a stand to form a bench comparator or into portable gauges such as air gap gauges.

The dial gauge head shown in fig. 4.12(b) as part of the height gauge may be replaced by an

Fig. 4.12. (a) Height setting micrometer.

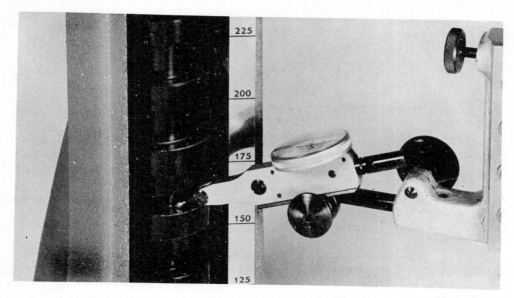

Fig. 4.12. (b) Detail of indicator probe reading on micrometer step. (By courtesy of Verdict Gauge (Sales) Ltd.) *Note:* **Air gauges** or electrical comparators may be used to replace test indicator if higher magnifications are required.

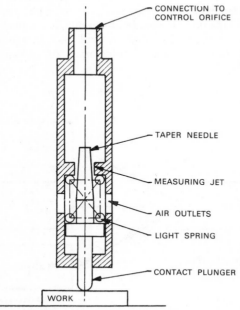

CONNECTION TO CONTROL ORIFICE

TAPER NEEDLE

MEASURING JET

AIR OUTLETS

LIGHT SPRING

CONTACT PLUNGER

WORK

Fig. 4.13. Plunger type of measuring jet for air gauge.

air gauge measuring jet as shown in fig. 4.14. Pressure on the ball-ended measuring probe, or stylus, flexes the thin portion of the measuring head body, thus opening or closing the measuring jet.

As long as certain precautions are observed, air gauges are an extremely useful form of measuring device for repetitive inspection and precision toolroom work, and can be built into a machine for process control during manufacture. The supply pressure must be kept constant by the use of sensitive regulators, the air must be filtered to avoid particles of dirt blocking the jets, and great care must be taken to ensure that the jets are not damaged. The manometer-tube types lend themselves readily to selective assembly work, as the scale can be replaced by coloured zones corresponding to the size grades used. This method, in conjunction with an air plug gauge, is used to grade cylinder bores in motor car engines.

87

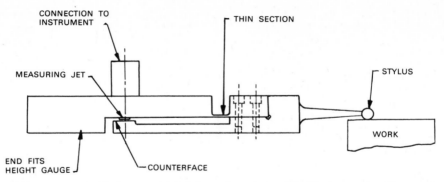

Fig. 4.14. Measuring jet suitable for use with height setting micrometer. The measuring force on the stylus causes the thin section to deflect and vary the distance between the jet and the counterface.

ELECTRICAL COMPARATORS

Specific Objective: *The student should be able to explain, with the aid of diagrams, the principle of operation of electrical comparators.*

Changes in position of a moving element in a measuring system can be measured electrically in a number of ways. The position of a slider along a variable resistance can be determined by measuring the change in resistance; the distance between a pair of capacitor plates can be determined by measuring the capacitance; the movement of an armature in the magnetic field of a pair of coils can be determined by measuring the change in inductance in the circuit. It is the inductive method which is most used in the field of metrology.

At this stage it is worthwhile examining the elements in a measuring system in the terms which will be used in later units in the field of instrumentation and control. All measuring systems consist of three units as shown in fig. 4.15.

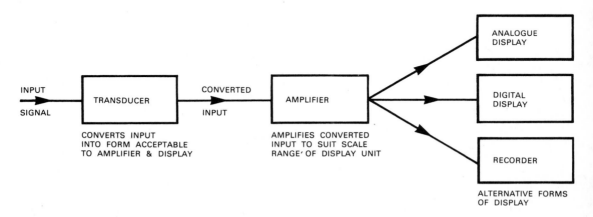

Fig. 4.15. Block diagram of measuring system.

(a) *Transducer*

The transducer unit is used to change the form of the input into a type of signal which can be handled by the rest of the system. Thus in an electrical measuring system for the measurement of small displacements, such as is used in the field of metrology, the function of the transducer is to convert the linear movement of a plunger into an electrical signal.

(b) *Amplifier*

Usually the output signal from the transducer is very small and must be amplified. In an electrical system the amplifier is electrical and while it is not within the field of this work to discuss the principles of electrical amplifiers it must be appreciated by the technician that the amplifier must be matched to the rest of the system. Assume that the transducer produces a voltage signal which is proportional to the distance moved by the plunger and this output is 2 millivolts per 0.01 mm movement. If the total movement of the plunger is to be ± 0.1 mm it means that the output voltage signal from the mid-position will be ± 20 mv. If now the instrument from which the reading is taken is a voltmeter which requires ± 10 V to give full-scale deflection from the mid-position, then the amplification must be:

$$\text{amplification} = \frac{\pm 10 \text{ V} \times 1000}{\pm 20 \text{ V}}$$
$$= \underline{500}$$

Thus an amplification, or gain, of $500\times$ is required if the whole of the pointer movement on the voltmeter is to be used in the measuring system.

(c) *Display unit*

The reading of the instrument may be taken in a number of ways. An *analogue scale* may be used, this usually being a pointer moving over a scale and is the most common form of display in measuring equipment. An alternative which is becoming increasingly used is the *digital reading* in which the reading is presented as a numerical value in figures. Students are probably familiar with these terms as they are used to describe clocks and watches, those with hands being analogue reading and those giving the time in illuminated numerical form being digital. Digital readings are less likely to be misread but it is not possible to interpolate the readings as can be done by estimating the position of the pointer between the marks on an analogue scale. Digital information is more acceptable to computers and if the output of the device is to be fed into a computer then a digital output can be fed direct into the computer while, in most cases an analogue output must be read by the operator and punched into the computer as a second operation.

A third form of display unit is the *recorder* which is used where dynamic measurements are being made and it is necessary to make a continuous record of the plunger or stylus movements as it traverses the work. Typical examples are found in the measurement of surface texture, roundness and errors in gears.

All measuring systems contain these three elements in one form or another and while the mechanical technician need not have a detailed knowledge of how they work he should be able to identify them and know their purpose. For example, in a dial gauge the transducer is the rack and pinion, which converts the linear movement of the plunger into a rotary motion as required by the amplifier, which is in the form of a compound gear train. To this is attached the pointer which moves over a circular scale to give an analogue read out.

In this section the student is concerned with the type of transducer that is used in electrical measuring systems. As has been stated, in measurements of the type encountered in the field with which this work is concerned the most common is the inductive type in the form of the *linear variable differential transformer*. The principle of this device is shown in fig 4.16. An armature is attached to the measuring plunger and around it is passed a coil. If a high frequency current is passed through this coil, magnetic fields are set up in the armature and these fields induce currents in two other coils wound on the same axis. These are connected so

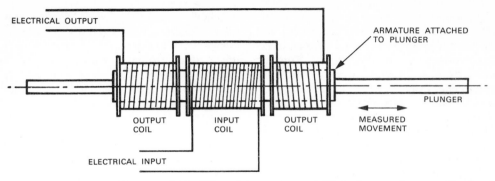

Fig. 4.16. Block diagram of linear variable differential transformer. This is the transducer used in electrical comparators.

that when the armature is in the mid-position their outputs are equal and opposite and cancel each other out. If the armature is moved to the right the output of one coil is greater than that of the other and a positive signal is fed to the amplifier. If the armature moves to the left the opposite occurs and a negative signal is obtained. Thus plus or minus readings can be shown on the display and, by switching the signal, it can be reversed.

These transducers can be made very small and, being connected to the amplifier and readout by flexible wires, they can be built into multi-gauging units. This has had an interesting effect on the type of display units used. It was found that for multi-gauging units using air gauges the water column manometers used for readouts could be conveniently grouped together and some modern electrical units use strip reading instruments which look like water columns.

If two transducers are placed opposite each other to measure the thickness of a component, their outputs can be fed into the same system and added. Their total reading is then the error in the thickness of the part and if the part is displaced it increases the output from one transducer and reduces it from the other by the same amount. Thus positioning of the gauge is not too critical. This technique is used on a modern comparator used for checking gauge

blocks at very high magnifications where the blocks are supported on a platen, through which one transducer protrudes from the underside, the other coming down from the top. The instrument is set to the reference gauge and then a reading is taken on the gauge to be tested. Slight vertical displacements of either gauge do not affect the measurement and the wringing of the gauge to the table is not as critical as with single transducer instruments.

Other advantages of electrical comparators are that both magnification and zero readings can be electrically adjusted. Thus where high magnifications are to be used the instrument can be initially set at low magnification, switched to high magnification and the zero adjusted by turning a knob on the readout which biases the reading one way or the other as necessary to give the fine zero adjustment.

It is interesting to compare electrical comparators with air gauges. Both use small transducers which can be built into measuring equipment as required. Both can be adapted to thickness measurement where displacement of the work is compensated for. Both connect the transducer to the amplifier using flexible connections and the amplifier can be remote from the transducer. Thus, both lend themselves to multi-gauging units. In both, the magnification can be easily changed, in one case by changing the electrical amplification and in the other by changing the area of the control orifice. Both lend themselves to operating recording devices and are used for dynamic measurements. Of the two, electrical devices are more readily adapted to the use of digital readout and thence to direct input to computer devices for complex work such as roundness testing, or where a printout is required of the results.

OPTICAL COMPARATORS
The TEC unit around which this work is written calls for optical comparators to be discussed at this stage. Modern optical comparators bear so much resemblance to other types of optical equipment to be discussed at a later stage that the author has decided to include them in that section and details will be found on page 104.

OPTICAL METHODS OF MEASUREMENT

OPTICAL PRINCIPLES

1. *Refraction*
If a ray of light passes from a less dense to a more dense material, e.g., from air to glass, it is 'bent' or *refracted* towards the more dense material, as show in fig. 4.17. As it passes from the more dense to the less dense material it is again refracted as shown. If we pass the ray through a prism it is refracted as shown in fig. 4.18.

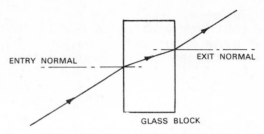

Fig. 4.17. Refraction. Note that the light is bent towards the normal on entry and away from the normal on leaving the glass.

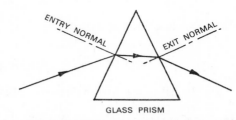

Fig. 4.18. Refraction of light through a glass prism.

A series of prisms can be arranged as shown in fig. 4.19 to bring all the light rays to a point, or *focus*. In practice this series of prisms is made in one piece and called a *lens*, and the distance from the focus to the lens is called the *focal length*.

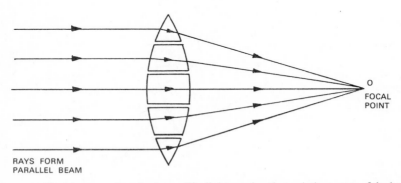

Fig. 4.19. A lens considered as a series of prisms. The light passing through the centre of the lens is not refracted.

91

2. Reflection

If a ray of light strikes a mirror at an angle θ to the normal to the mirror, the reflected ray will also make the angle θ with the normal, as shown in fig. 4.20.

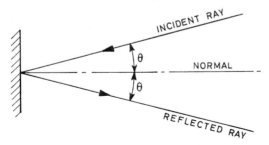

Fig. 4.20. Reflection from a plane surface.

The angle of incidence is equal to the angle of reflection.

If now the reflector is turned through a small angle δθ, the normal also turns through δθ, making the angle of incidence (θ + δθ). The reflected angle will also be (θ + δθ), as shown in fig. 4.21.

∴ Total angle between incident

$$\text{and reflected ray} = 2(\theta + \delta\theta)$$
$$= 2\theta + 2\delta\theta$$

Original angle between incident

$$\text{and reflected ray} = 2\theta$$
$$\therefore \text{Angle turned by reflected ray} = 2\delta\theta$$

This is the principle of the *optical lever*.

If a mirror is turned through a small angle the reflected ray turns through twice that angle.

This principle is used in optical instruments to give a 'free' doubling of the magnification.

3. Collimation and De-Collimation

On referring again to fig. 4.19 it is obvious that if we reverse the situation, with a suitable lens, and place a point of light at the focal point, O, of the lens it will be projected as a parallel beam of light. Such a lens is called a *collimating lens*. If the parallel beam of light projected from a collimating lens strikes a similar lens it is refocused or *de-collimated* at point O_1 as shown in fig. 4.22.

It is upon these principles that most of the optical equipment used in engineering measurements are based.

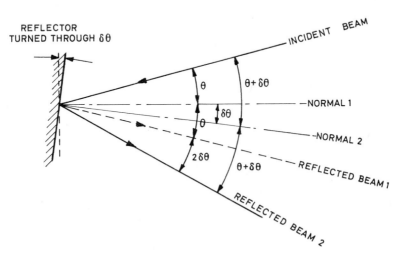

Fig. 4.21. Reflection from a plane surface turned through angle δθ.

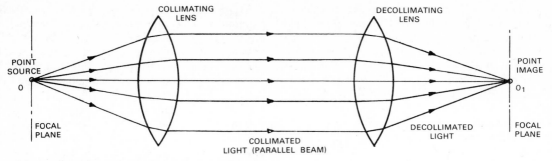

Fig. 4.22. Collimation and de-collimation of light.

THE OPTICAL PROJECTOR

General Objective: *The student should recognise the principle and use of the optical projector.*

Specific Objective: *The student should be able to describe and explain, with the aid of diagrams, the design of the optical projector.*

If an object is placed behind a lens of the type shown in fig 4.19 and illuminated from behind by a parallel beam of light its image will be *projected* by the lens, the extreme rays travelling the paths shown in fig. 4.23. By the properties of similar triangles:

$$\frac{\text{size of image}}{\text{size of object}} = \frac{l}{f}$$

Fig. 4.23. Light ray path in an optical projector. 'l' is changed to adjust the magnification. 'l_1' is changed to adjust the focus.

where l = distance between lens and screen;
f = focal length of lens.

This is the principle of the optical projector. Its basic elements are:

1. A lamphouse which projects a parallel beam of light.
2. A mounting for the work or workstage.
3. A projection lens.
4. A screen.

In order to adjust the magnification it is necessary to adjust the distance between the lens and the screen and thus change the ratio l/f; alternatively the magnification can be changed by changing the lens to one of a different focal length, 'f'. In order to focus the image, it is necessary to adjust the distance between the object and the lens l_1, so that the rays passing through the centre of the lens and those passing through the focal point coincide

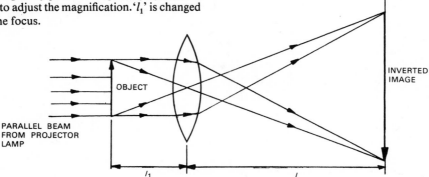

93

at the screen. In the case of simple slide and 'home movie' projectors, this is done by moving the lens. However, it has already been shown that moving the lens will change the magnification, so, in the case of precision projectors for engineering measurements, it is necessary to adjust the focus by *moving the object*.

Such an instrument, having all these features and adjustments, is shown in fig. 4.24. More modern instruments exhibit the same features but, by the use of mirrors and prisms in the optical system the ray path is 'bent' to give a more convenient unit, and project the image on a screen close to the operator and work stage so that all operations can be carried out from one position.

Specific Objective: *The student should be able to explain why a collimated beam of light is necessary in optical projection.*

In listing the essentials of an optical projector it was stated that the lamphouse must project a parallel beam of light. If a ray diagram similar to fig. 4.23 is drawn as in fig. 4.25, but with a diverging light beam, it can be seen that the apparent focal length of the lens, and hence the magnification, is effectively changed. As many projectors are of fixed magnification it follows that incorrect focusing of the illuminating beam will result in a change in the magnification and incorrect measurements will result. In fact, fig. 4.25 is an over simplification of the situation. The boundary rays only are shown and will cross at the new apparent focal point, but all other rays nearer to the centre of the lens, will cross the optical axis at different positions nearer to the true focal point. Thus a sharply focused image will be unattainable and accurate measurements on the image will be difficult. If an aproximate focus is obtained, the magnification will be incorrect.

Applications of the optical projector

Specific Objective: *The student should be able to describe, with the aid of diagrams, the principle of operations of the optical projector.*

The main use of an optical projector is to enable small features to be 'blown up' into an enlarged

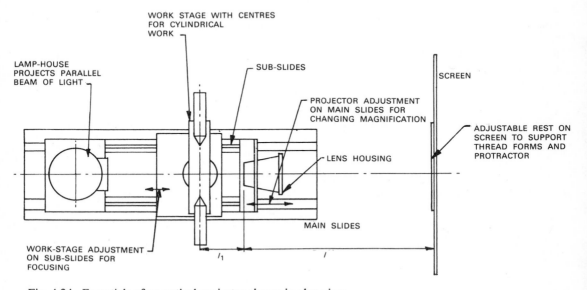

Fig. 4.24. Essentials of an optical projector shown in plan view.

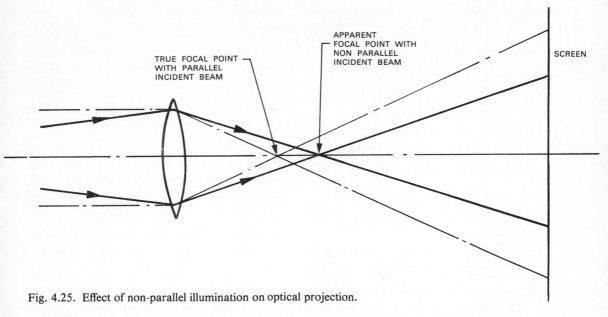

TRUE FOCAL POINT
WITH PARALLEL
INCIDENT BEAM

APPARENT
FOCAL POINT WITH
NON PARALLEL
INCIDENT BEAM

SCREEN

Fig. 4.25. Effect of non-parallel illumination on optical projection.

image so that they can be measured. The accuracy of the measurement depends on the level of magnification and the method of measurement being used. Thus it is necessary to check the magnification regularly. A simple method of doing this is to project an object of known size and measure the image. A small calibrated cylinder 2 mm diameter is suitable, the image being measured with a rule. This may seem crude, but if a standard magnification is used, 25× or 50× being common, then the error in the actual measurement is divided by the magnification. Thus if the 50× magnification is used, the image should be 100 mm across. If the measurement is carried out to within 0.5 mm, then the accuracy of determination will be within 0.005 mm. In any case, if this is the method to be used to measure the image of the object under test, then the same accuracy applies. Unfortunately this only checks the magnification over one area of the screen. A better method is to project a transparency ruled off into small squares. By measuring some of the projected squares the magnification can be checked, and by observation any distortion of the image over the whole field can be seen.

Having checked the magnification, the workpiece to be measured can be set up, properly focused and measurement can be carried out. There are three methods used generally for checking work by projection:

1. By comparison with a carefully drawn master template.
2. By measuring the image at the screen.
3. By micrometer-controlled movements of the object.

Use of templates
If an object is to be projected at 25× magnification, then the template must be carefully drawn at that magnification. Here it is necessary to state a simple truth which is often overlooked. If a length of 2 mm is magnified 25×, it becomes 50 mm. If an angle of 30° is magnified 25×, it is still 30°. All that happens is that the lengths of the lines forming that angle are magnified. Thus it is not good enough to carefully draw lengths to the required magnification and then set off the angles with an adjustable set square or the head on a drafting machine. Angles should be set off by calculating

95

coordinate dimensions and carefully laying them out at right angles. For example, if an angle of 25° is to be drawn through a point the procedure should be:

1. Draw a line vertically through the point and measure off a convenient length, say, 200 mm.
2. Calculate $200 \tan 25° = 200 \text{ mm} \times 0.446\,307$
$$= 93.26 \text{ mm}$$
3. Through the 200 mm point on the vertical line draw a horizontal line, and measure off 93.26 mm as acurately as possible.
4. Draw through this point and the original.

Note that it will not be possible to set off the 200 mm and 93.26 mm exactly, but the error incurred will be much less than that caused by trying to draw the angle directly.

Having drawn the template, it must now be aligned with the image. If the template is drawn on aluminium it can be rested on an adjustable ledge, attached to the screen as shown in fig. 4.24. This ledge should have been adjusted to align with the datum feature of the image.

Note that in many cases, the template will have been drawn showing *not only* the basic size of the workpiece *but also* the tolerance zone at the correct magnification. In these cases it is only necessary to ensure that the whole of the image falls within the tolerance zone.

Direct measurement of images
This can be done either by actually measuring the image at the screen using a rule or by putting paper on the screen and, using a sharp hard pencil and straight edges, drawing over the features of the image to be measured. The drawing can then be taken away to be measured on a drawing board, which is more convenient.

A particular application of this method is in the assessment of surface finish where an image of the trace is projected, drawn over and taken away to be analysed. This method has largely been superseded by more modern methods of assessment of surface texture and measurement of the image on the screen has similarly been outmoded in modern projectors by controlling the movement of the object.

Controlled movement of the object
Consider a screen having marked on it two lines at 90° to each other, against which the image may be viewed. If the workstage of the projector can move on slideways at right angles to the optical axis, and this motion is imparted by a large drum micrometer, then dimensions can be measured by moving the object rather than measuring the image. Imagine a component having a groove machined in it. The workstage is adjusted until the image of one face of the groove coincides with the datum line on the screen and the micrometer reading is noted. The micrometer is adjusted to move the workstage until the image of the other face of the groove coincides with the datum line and the micrometer reading again noted. The difference in the two readings is the width of the groove.

If the depth of the groove is required, then a micrometer controlled movement at 90° to the first is necessary, the procedure being the same.

If it is necessary to measure angles, then the datum lines on the screen must have the facility of being rotated and this rotation measured by an angular scale which can be read to the required degree of accuracy. What has been described here is the layout of the modern optical projector or toolmaker's microscope. One motion which has not been mentioned is a rotary motion of the workstage in the plane of the linear motions for setting purposes. These three motions are usually in the horizontal plane as shown in fig 4.26(a). To use such an instrument to make a measurement the procedure is:

1. Set up the work on the workstage and focus to give a sharp image.
2. Adjust the angular scale on the screen to read 0°. If the instrument is correctly adjusted this means that the datum lines on the screen are parallel with the linear movements of the workstage as projected.
3. Rotate the workstage until the datum surface of the workpiece image is parallel to the datum lines on the screen.
4. Make linear measurements using the micrometer adjustments of the workstage.

5. Make angular measurements by rotating the datum lines on the screen to coincide with angled faces on the image. The views on the screen in making the measurements A and B (fig. 4.26(a)) are shown in fig 4.26(b).

SCREW THREAD PROJECTION

Specific Objective: *The student should be able to describe, with the aid of diagrams, the adjustments necessary to project helical forms.*

A common use for the optical projector is the projection of screw threads to check the accuracy of form either on high-class commercial threads or on screw plug gauges. Accurate gear forms will only be produced on a hobbing machine if the hob form is correct. In all these cases the form to be projected is not a parallel groove cut in a cylinder but is part of a helix cut around and along the cylinder. If a helical component is projected with a projector as shown in fig. 4.27(a) it will be difficult to produce a sharp image of the form because the helix angle interferes with the passage of the light rays. Many projectors overcome this problem by allowing the workstage to be swung through the helix angle of the form to allow the rays to pass straight through. Unfortunately the *length* of thread projected is shortened, the

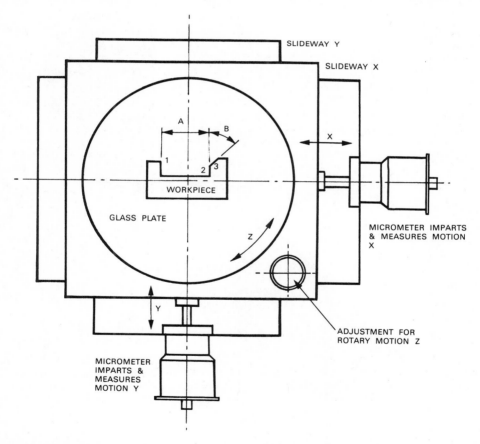

Fig. 4.26. (a) Workstage of toolmakers' microscope showing adjustments.

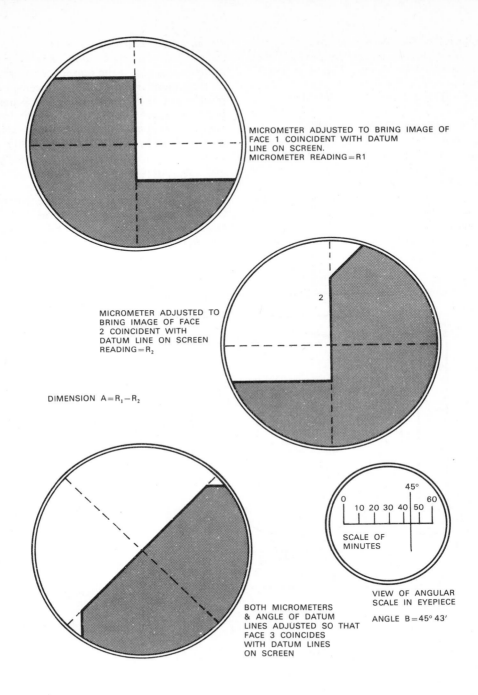

MICROMETER ADJUSTED TO BRING IMAGE OF FACE 1 COINCIDENT WITH DATUM LINE ON SCREEN.
MICROMETER READING = R1

MICROMETER ADJUSTED TO BRING IMAGE OF FACE 2 COINCIDENT WITH DATUM LINE ON SCREEN
READING = R_2

DIMENSION $A = R_1 - R_2$

BOTH MICROMETERS & ANGLE OF DATUM LINES ADJUSTED SO THAT FACE 3 COINCIDES WITH DATUM LINES ON SCREEN

VIEW OF ANGULAR SCALE IN EYEPIECE

ANGLE B = 45° 43′

Fig. 4.26. (b) Views on screen during measurement of workstage shown in fig. 4.26(a).

98

depth of thread is not affected and the flank angles projected will be slightly less than they actually are.

*To project a helical form correctly the work axis, lens and screen **should always be parallel.***

This means that if the rays of light from the lamphouse are to pass through the thread without interference, it is necessary to swing the lamphouse through the helix angle.

On most single-start threads where the helix angle is small, the effect of rotating the work is negligible but in the case of multi-start components such as hobs the error can be significant, and to avoid the possibility of such errors it is always the lamphouse which should be swung. This effect of foreshortening when the workstage is swung through the helix angle is shown in fig. 4.27(b).

The actual measurement of the angles of a thread can be carried out using a shadow protractor mounted on the straight edge on the screen with a simple projector. This is shown in fig. 4.28, while fig. 4.29 shows the image on the screen of a more modern projector with the datum lines inclined to coincide with the image of the thread flanks.

ANGULAR MEASUREMENT

General Objective: *The student should recognise the principles of angular measurement by optical methods.*

In the work on measurement in level 2[1]
[1] *Technician Manufacturing Technology 2* (Cassell), pp. 249/72.

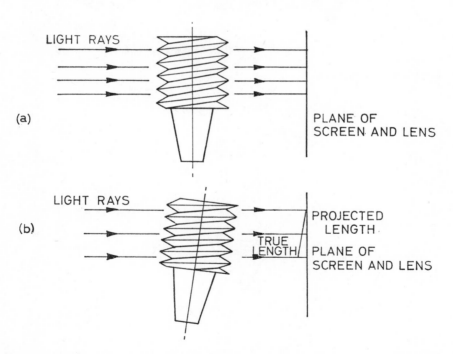

Fig. 4.27. (a) Projection normal to thread axis causes interference, and (b) Turning thread through helix angle avoids interference, fore-shortens pitch and distorts profile.

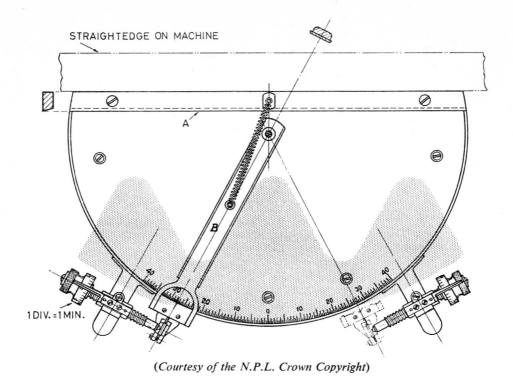

STRAIGHTEDGE ON MACHINE

A

B

1 DIV. = 1 MIN.

(Courtesy of the N.P.L. Crown Copyright)

Fig. 4.28. Flank angle protractor.

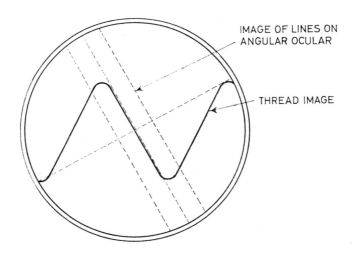

IMAGE OF LINES ON
ANGULAR OCULAR

THREAD IMAGE

Fig. 4.29. Image of screw thread on screw of toolmakers' microscope.

100

mechanical methods such as the vernier protractor and sine bar were considered along with measurements over rollers and balls from which angles could be derived. In many cases these methods are slow and cumbersome and the determination of angles can be carried more rapidly and to a higher degree of precision using optical equipment, and in particular using an optical device called an *Angle Dekkor*. The author recomends to any student using an Angle Dekkor that three questions should be asked:

(a) How else could the measurement be made?
(b) Would it be as quick by that method?
(c) Would it be as precise by that method?

In almost all cases, whatever the answer to the first question, the answer to the second and third will be 'no' in both cases.

The Angle Dekkor

Specific Objective: *The student should be able to describe, with the aid of diagrams, the principle of operation of the Angle Dekkor.*

Earlier in·this section, various optical principles used in measurement were discussed. In the case of the Angle Dekkor we shall refer to the principles of collimation and de-collimation, and reflection. The Angle Dekkor is a form of *Auto-Collimator* in which the collimation and de-collimation are carried out by the same lens. This principle is illustrated in fig. 4.30(a) which shows a point source of light, 'O' in the focal plane of a collimating lens. It is projected as a parallel beam of light, but instead of passing through a de-collimating lens it strikes a plane reflector set at 90° to the optical axis. It is therefore reflected back along its own path and refocused at the source 'O' by the same lens through which it was projected; this is called *auto-collimation*. If the reflector is now tilted

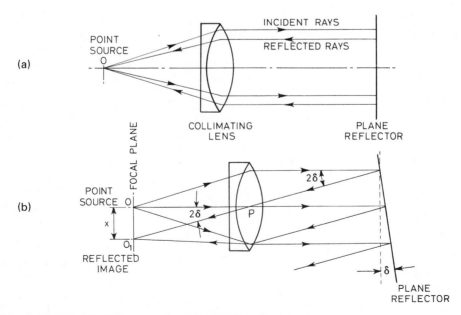

Fig. 4.30. (a) Point source of light in focal plane of a collimating lens.
(b) Projection of a point source being reflected from an inclined reflector.

101

through a small angle 'δ' the reflected beam will turn through 2δ, as described in the section on the reflection principle, and will be refocused to produce an image of the source at O_1, a distance x from the source O as shown in fig. 4.30(b). Thus the distance, between the image and the source, x, is a measure of the angular displacement of the reflector, and from fig. 4.30(b):

In triangle POO_1

$$\frac{OO_1}{OP} = \tan 2\delta$$

$$x = OO_1 = OP \tan 2\delta$$

OP is the focal length of the lens and as δ is small:

$$\tan 2\delta = 2\delta \text{ radians}$$
therefore $\qquad x = 2\delta f \text{ mm}$

There are important points about this expression which are not immediately apparent. These are:

(a) The distance between lens and reflector has no effect on the separation between source and reflected image.

(b) For high sensitivity, i.e. a large value of x for a small angular displacement δ, a long focal length of lens is needed.

(c) Although the distance of the reflector from the lens does not affect the reading x, if, at a given value of δ the reflector is moved too far back all the reflected rays will miss the lens and an image will not be formed. Thus for a wide range of readings, the reflector should be as close as possible to the lens. This is particularly important in the case of optical comparators which, as will be shown, are a form of Angle Dekkor modified for linear rather than angular measurement.

It is not practical to project the image of a point source of light as has been suggested. In the case of the Angle Dekkor a glass screen in the focal plane of a collimating lens has engraved on its underside a scale of minutes.[1] It

[1] It is in fact engraved as a mirror image so that after reflection it can be read normally.

is illuminated via a small prism and is projected by the collimating lens as a parallel beam of light, strikes a plane reflector and is refocused by the lens to appear as a reflected image on the screen. The image falls, not across a simple datum line but across a similar fixed scale at right angles to the illuminated image. Thus the reading on the illuminated scale measures angular deviations from one axis at 90° to the optical axis, and the reading on the fixed scale gives the deviation about an axis mutually at right angles to the other two.

This feature enables angular errors in two planes to be dealt with or, more important, to ensure that the readings on a setting master and on the work are the same in one plane, the error being read in the other. Thus, induced compound-angle errors are avoided.

The optical system and the view in the eyepiece are shown in fig. 4.31. The physical features simply consist of a lapped flat and reflective base above which the optical details are mounted in a tube on an adjustable bracket.

The Angle Dekkor is extremely useful for a wide range of angular measurements at short distances. It therefore finds its application in toolroom-type inspection. Readings direct to 1′ over a range of 50′ can be taken and, by estimation, readings down to about 0·2′ are possible.

The Angle Dekkor is essentially an angle comparator and as such it must be set to a calibrated master gauge in just the same way that a comparator for linear measurement is set to gauge blocks.

Specific Objective: *The student should be able to explain how the Angle Dekkor is used in conjunction with combination angle gauges for the measurement of a given angle.*

The most common setting gauges used with an Angle Dekkor are *Combination Angle Gauges*. These are wedge-shaped blocks of hardened plain carbon steel which have been stabilised, and their working faces lapped to the stated angle and to a wringing finish. They can thus be wrung together, as can slip gauges, to make up

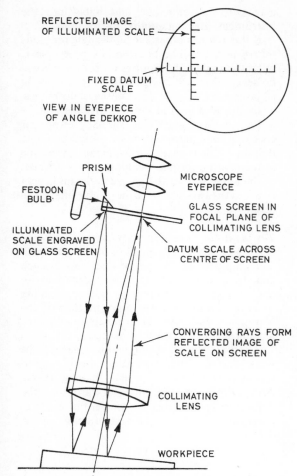

Fig. 4.31. Optical system of an Angle Dekkor.

angles. There is one important difference however; being angles they can be subtracted as well as added. Thus 9° and 27° blocks can be combined to form an angle of 36°. If now the 9° block is reversed, as shown in fig. 4.32, an angle of 18° is formed. A full set of combination angle gauges consists of the following pieces:

Degrees: 1° 3° 9° 27° 41°
Minutes: 1′ 3′ 9′ 27′
Decimal
Minutes: 0.05′ 0.1′ 0.3′ 0.5′
A square block

By suitably combining these gauges, any angle between 0° and 90° can be made up in intervals of 0.05′ or 3 seconds. For most work with an Angle Dekkor, which has a scale of 50′ the full set is not required and a smaller set consisting of the degree gauges and the square block is suitable.

Most work pieces do not have a suitable reflective surface off which readings can be taken direct and plane steel reflectors can be purchased, whose reflective surface is lapped accurately parallel with its back face. To take a reading off the workpiece, this reflector is simply placed on the work to give a clear image in the instrument eyepiece. If such a reflector is not available, a small gauge block about 5 mm thick is perfectly adequate.

Consider now a block 30 mm long ground to an angle of 5° 10′, the thickness at the small end being 10 mm and the limits on the angle being specified as ±2′. The block would probably have been ground using a sine table on a surface grinding machine, but it is too short to be reasonably checked on a sine bar as the length, along which it could be checked using a dial gauge, is not adequate. In order to check the angle using an Angle Dekkor the procedure is as follows:

1. Make up an angle of 10° using combination angle gauges. The gauges required would be 9° − 3° − 1° = 5°.

2. Place these gauges on the table of the Angle Dekkor and roughly square them up. This can be done easily by setting them against a square registered off the edge of the table.

3. Adjust the Angle Dekkor in the vertical plane until the illuminated scale appears in the

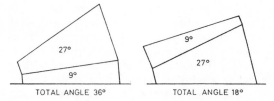

Fig. 4.32. Addition and subtraction of combination angle gauges.

eyepiece. It will not be in the centre of the eyepiece but this can be adjusted by slightly rotating the gauges in the horizontal plane.

4. When the instrument is set satisfactorily note the readings on both the illuminated and the fixed scale.

5. Remove the combination angle gauges and replace them with the workpiece upon which a slip gauge is resting. Adjust the workpiece in the horizontal plane until *the same reading is obtained on the fixed scale*. Now note the reading on the illuminated scale. The difference between this reading and that obtained on the combination angle gauges is the difference in their angles and should, in this case, be 10′.

Note the importance of taking the readings on the illuminated scale with the same reading on the fixed scale. If this is not done an error in the result will occur due to the compound angle in the setting of the two pieces.

The above procedure is used in all measurements of angles with the Angle Dekkor. It can be used to check conical parts such as taper plug gauges in which case the slip gauge can be held in place on the plug gauge with plasticine to make for easy handling. Both pieces must be thoroughly cleaned after the work is finished.

Testing right angles

We have seen that an auto-collimator or Angle Dekkor projects the image of a target wire or scale as a parallel beam of light. This beam of light consists of an infinite number of parallel rays, each of which is itself a minute image of the scale.

If we project the beam from an Angle Dekkor at an internal right angle, for instance a vee block as shown in fig. 4.33, the rays striking reflector (A) directly are reflected across to reflector (B) and back into the instrument, thus forming a reflected image of the scale. Due to the additional reflecting surface involved this will be a mirror image of the scale. Similarly the rays striking reflector (B) directly are reflected across to reflector (A) and back into the instrument, forming another mirror image.

If the angle is a perfect right angle only one mirror image will be observed in the eyepiece, but if the angle is in error *two* mirror images will be seen, as shown in fig. 4.33(b). The error in squareness is *half* the offset of the scale images, again due to a built-in reversal process.

Fig. 4.33(c) shows a similar set-up for checking an external angle. It should be noted that in neither case is the alignment of the instrument important. As long as it is directed at the angle so that the projected beam is split between the two reflectors the double image will result.

Testing a 180° angle (parallelism)

If a workpiece is to be checked for parallelism it can simply be measured for thickness at each end. If, however, the error is required as an angle it can be determined by taking a reading by Angle Dekkor on the table and then on the workpiece, the difference in readings giving the angular error in parallelism directly.

Alternatively, a reading can be taken on the work surface, the work then being turned through 180° and a repeat reading taken. Again, due to the reversal process, the difference in readings is *double* the error in parallelism. This is shown in principle in fig. 4.34.

THE OPTICAL COMPARATOR

Specific Objective: *The student should be able to explain, with the aid of sketches, the principles of operation of optical comparators.*

If an Angle Dekkor is used in conjunction with a reflector whose angular position depends on the position of a measuring plunger as shown in fig. 4.35, when the plunger is raised a small distance h and the reflector is pivoted at a distance y from the centre of the plunger, then the reflector will be turned through a small angle δ:

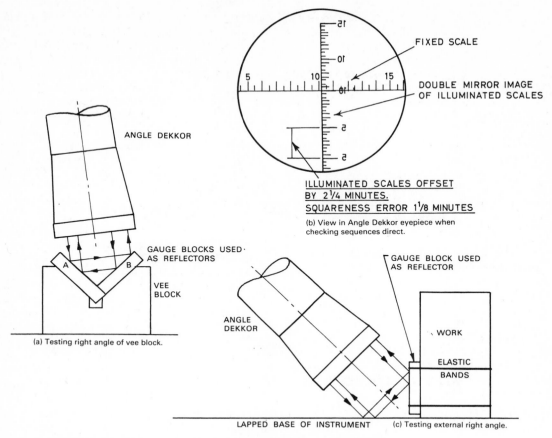

(b) View in Angle Dekkor eyepiece when checking sequences direct.

(a) Testing right angle of vee block.

(c) Testing external right angle.

Fig. 4.33. Direct measurement of errors in right angles using an Angle Dekkor.

$$\text{Angle } \delta = \frac{h}{y} \text{ radians}$$

Applying the auto-collimator principle, the displacement x of the scale image is given by:

$$x = 2\delta f$$

where f = the focal length of the collimating lens.

Therefore

$$x = \frac{2h}{y}f$$

The magnification is the ratio of the scale movement x to the plunger movement h, i.e.,

Magnification

$$= \frac{x}{h} = \frac{2hf}{y} \times \frac{1}{h}$$

$$= \frac{2f}{y}$$

This is increased further by the magnification of the eyepiece and thus:

Overall magnification

$$= \frac{2f}{y} \times \text{eyepiece magnification}$$

105

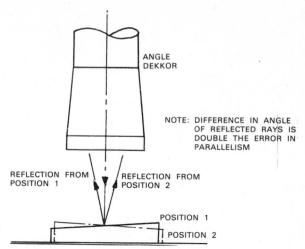

ANGLE
DEKKOR

NOTE: DIFFERENCE IN ANGLE
OF REFLECTED RAYS IS
DOUBLE THE ERROR IN
PARALLELISM

REFLECTION FROM
POSITION 1

REFLECTION FROM
POSITION 2

POSITION 1

POSITION 2

Fig. 4.34. Use of the reversal method in testing parallelism.

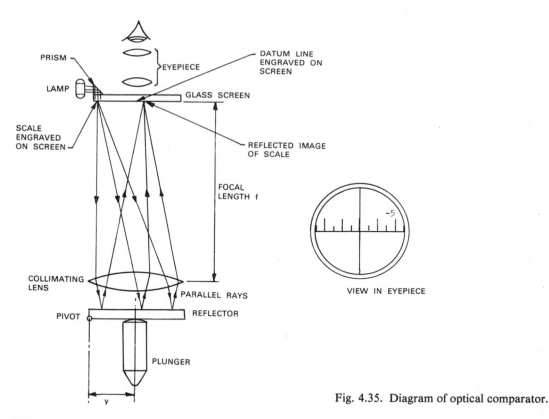

PRISM

EYEPIECE

LAMP

DATUM LINE
ENGRAVED ON
SCREEN

GLASS SCREEN

SCALE
ENGRAVED
ON SCREEN

REFLECTED IMAGE
OF SCALE

FOCAL
LENGTH f

COLLIMATING
LENS

PARALLEL RAYS

PIVOT

REFLECTOR

PLUNGER

y

VIEW IN EYEPIECE

Fig. 4.35. Diagram of optical comparator.

106

In practice the reflector is built into the instrument and the scale is in units of linear movement of the plunger. A prism is built into the instrument to give a horizontal eyepiece as shown in fig. 4.36. To avoid eye strain, most instruments of this type do not use an eyepiece but project the image of the scale on a screen which can be easily read, even under poor lighting conditions.

An advantage of optical comparators over most other types is the small number of moving parts — simply the plunger and the reflector. There is very little to go wrong with them other than bulb replacement

It will now be appreciated by the reader why the author has included this section on optical comparators with the Angle Dekkor. The two instruments are basically the same, the optical comparator being used for measurement of length by comparison with length standards and the Angle Dekkor for the measurement of angles by comparison with standards of angle.

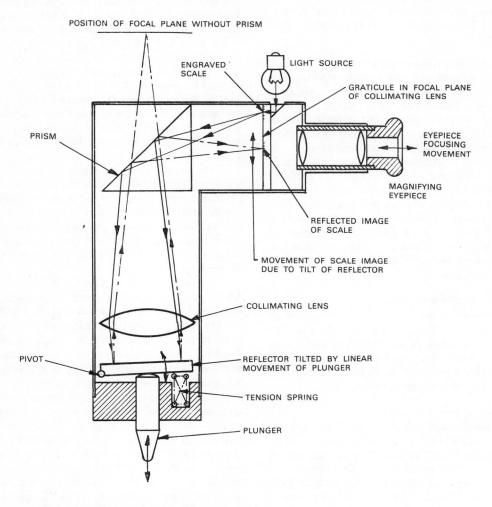

Fig. 4.36. Movement of optical comparator.

It is, then, appropriate at this stage to briefly compare the characteristics of the different types of comparator.

1. *Mechanical comparators*
These are probably the most used type of comparator. They are completely self-contained and require no external power supply but the transducer cannot be separated from the amplifier and display units. They do not, therefore, lend themselves to building into multi-gauging units but are most useful for standing by a machine to be used for checking a critical dimension of a part as it comes off the machine.

2. *Pneumatic comparators*
The transducer unit can be separated from the amplifier and display units and connected to them by flexible plastic tubes. As the measuring jets can be small units, these two facts, combine to enable this type of instrument to be used in multi-gauging units and instruments where the measuring head must move. They also can be easily adapted to operate a recorder. Magnification can be changed and adjusted by varying the control jet. Open jets tend to clean foreign matter off the work and, when used with an air plug gauge, a pneumatic gauging unit represents one of the few ways of measuring inside diameters to a high degree of precision easily and quickly.

3. *Electrical comparators*
These have almost all the advantages of air gauging with the exceptions that they are contact methods of measurement and do not provide a self cleansing facility. Generally they tend to be more stable and it is not necessary to check the magnification so frequently.

4. *Optical comparators*
Like mechanical comparators, these are one piece units and it is not possible to separate the transducer from the amplification and display systems. They are mainly used in metrology laboratories and standards rooms, their main advantages being the clarity of scale, ease of reading and few moving parts.

MEASUREMENT OF SURFACE ROUGHNESS

General Objective: *The student should recognise the principles of specification and measurement of surface finish*

The necessity to work to higher degrees of accuracy has brought with it the necessity to produce surfaces of better quality. It is of little use specifying a tolerance on the size of a part to the order of $\pm 0.000\,3$ mm if the local variations in height which constitute the surface finish are of the order $0.001\,5$ mm.

Furthermore, the nature of the surface may influence the functioning of the part. If a shaft is subject to reversals of load it becomes fatigued and its life can be greatly reduced. Failure due to fatigue often starts at the sharp root of a surface irregularity, and it follows that the better the surface finish the longer will be the fatigue life.

Wear is another problem confronting the engineer. The useful life of an assembly is often governed by the rate of wear of its component parts, and the rate of wear depends on the surface area in contact. A rough surface with large peaks and valleys will have less contact area than a smooth one and its rate of wear will be greater.

If the surface finish of a part is to be specified for a given function it must be possible to measure (assess is a better word) the surface quality and give it a numerical value. Visual assessment and assessment by touch are both used but these are subjective, i.e. they allow surfaces to be compared but only on the basis of an opinion as to which is the rougher.

A numerical assessment is better so that we can say that the surface to which the larger value is assigned is the rougher. This numerical assessment is usually obtained by means of an instrument in which a stylus is drawn slowly over the surface, and the stylus movements are measured to provide the numerical assessment. This method, however, raises a problem concerned with the geometry of surfaces.

Specific Objective: *The student should be able to describe, with the aid of a diagram, the three broad deviations of a surface from the true mean plane and state which of these deviations is considered in the term 'surface finish'.*

The overall profile of a cross-section through a machined surface is complex in its geometry and in order to measure the roughness it is first necessary to define the term and, having done so, to design a measuring instrument which only measures roughness and eliminates the effects of other errors from the perfect surface which do not contribute to the roughness.

Consider the surfaces shown in figs. 4.37, 4.38 and 4.39. Each is 200 mm long, and all have changes in height which are equal. If a small piece of metal, which we shall call a *shoe* or *skid*, about 10 mm long and having a radius of 100 mm, is drawn across a perfectly flat and smooth surface it will move in a perfect straight line.

The surface shown in fig. 4.37(a) has a deviation from the straight line of 0.005 mm and is depicted as a sine wave whose wavelength is equal to the length of the surface; the skid will move smoothly along it but will move up and down – as it does so following the error in form.

A similar thing happens to the skid in the case of fig. 4.37(b), but it moves up and down more often, the wavelength being shorter, and the surface could still be described as smooth, but not flat.

If, however, the wavelength is made extremely short, as shown in fig. 4.38, the skid would again move in a straight line simply riding on the peaks of the imperfections, and this surface could be described as flat but rough.

As in each case the heights of the imperfections are identical, it follows that the difference between roughness and non-flatness is not one of depth of imperfection but of spacing. In the past, various methods have been suggested to differentiate between roughness and non-flatness but the currently accepted definitions as set out in B.S. 1134 (1972) are as follows:

1. *Roughness*
Those irregularities arising from the cutting action of the tool producing that surface. These include feed marks and small tears produced when the chip is ruptured from the material.

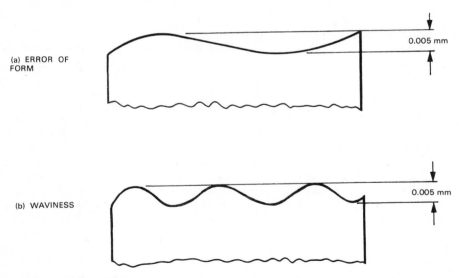

(a) ERROR OF FORM

0.005 mm

(b) WAVINESS

0.005 mm

Fig. 4.37. Two non-flat surfaces.

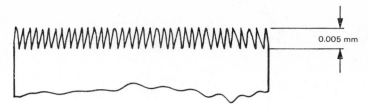

Fig. 4.38. Surface which is flat but rough.

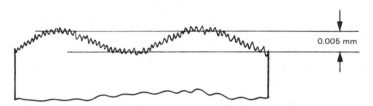

Fig. 4.39. Roughness superimposed upon non-flatness. An instrument measuring roughness must eliminate the waviness component.

2. *Waviness*

Those irregularities resulting from vibration and short wavelength deflections of the machine or workpiece during the machining process.

3. *Errors of form*

Those irregularities resulting from errors in machine tool geometry and alignments, deformation of the workpiece due to cutting forces or machine deformations due to the workpiece mass.

In practice, these irregularities in the surface are usually superimposed one upon the other as shown in fig. 4.39, and it is clear that if a stylus was drawn across the overall length of the surface and its movements measured relative to a perfect straight line, all three types of imperfections would be included. Any instrument used to assess the surface finish must only measure the roughness and must be designed to eliminate waviness and errors of form.

INSTRUMENTS FOR THE MEASUREMENT OF SURFACE FINISH

Specific Objective: *The student should be able to describe, with the aid of diagrams, the principles of construction and operation of the stylus-type of measuring instruments for the measurement of surface finish.*

In order to eliminate the effects of errors of form and waviness all instruments which measure the roughness of a surface by drawing a stylus[1] over it have certain features in common, the principles of which are shown in fig. 4.40, in the form of a block diagram.

A skid, having a large radius, is carried on an arm mounted on a pivot, A. The skid rests on

[1] Other methods, specifically optical methods, are not considered here.

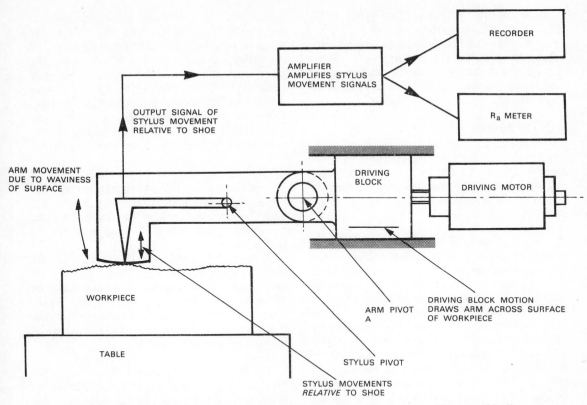

Fig. 4.40. Block diagram of stylus/shoe instrument for assessing surface texture. Compare this diagram with fig. 4.15.

the surface, and as the arm is drawn slowly across the surface, the skid follows the general profile of the work. A stylus is mounted within the arm and, due to the irregularities of the surface, moves *relative to the skid* – it is these movements of the stylus *relative to the skid* which represent the roughness of the surface and the shorter wavelength components of the waviness which are not eliminated by the skid. The movements of the stylus *relative to the skid* are amplified and used to operate a recorder from which a trace, representing a cross-section of the surface, may be obtained and analysed to give a numerical assessment of the surface. Alternatively the signals produced by the stylus movement may be used to operate a roughness meter which gives a numerical assessment of the surface direct.

The shorter wavelength components of the waviness, which are not eliminated by the skid following the profile of the surface, are excluded from the assessment by limiting the length of surface included in the assessment. This is discussed more fully on p. 116, but it should be noted that all instruments traverse a short length of surface as part of the elimination of waviness.

To summarise, it can be seen that all stylus-type instruments for the assessment of surface finish have the following features in common:

1. An arm, mounted on a pivot carrying a skid, which is drawn slowly across the surface for a short distance.

2. A stylus mounted within the arm which is caused to move relative to the skid by the roughness of the surface.

111

3. A means of amplifying the stylus movements relative to the skid.

4. A means of producing a magnified trace representing the surface, *and/or*

5. A meter to give a direct reading of the roughness of the surface as a numerical assessment.

COMPARISON OF TRACE AND SURFACE

Specific Objective: *The student should be able to sketch and explain the difference between the shape of the graphical results and the actual surface measured.*

A common feature of all recording instruments used in the assessment of surface texture is that the magnification of the vertical movement of the stylus is much greater than the magnification of the horizontal movement of the stylus across the surface. The vertical magnification may be up to 50 000× while the horizontal magnification rarely exceeds 100×.

The stylus used in recording instruments has a tip radius of 0.002 mm and an angle of 90° ± 5°. The author is often asked how a stylus of this form can get down into a valley having the profile shown at A in fig. 4.41; the answer is that it doesn't. The vertical magnification of this trace is 2000×, and the horizontal magnification only 20×. Thus the real profile of the valley can only be found by redrawing the trace with the horizontal distance between peaks B and C magnified 100× to give the same vertical and horizontal magnification. It must, therefore, be emphasised that a trace taken from a surface is by no means a true record, but is greatly distorted. There are two reasons for this:

1. If the trace in fig. 4.41 is redrawn with the horizontal magnification equal to the vertical magnification, the angle of the features will be so shallow that many features will not be recognisable.

2. If the horizontal magnification in fig. 4.41 was increased by 100× the trace, which is 10 cm long, would become extended to a length of 10 m — the width of a wide street. Again scanning the trace in order to discern features would be an impossible task — to say nothing of the cost in recorder paper.

To show how the features discussed on the previous pages are incorporated into working instruments, two will be considered, one mechanical and the other electronic.

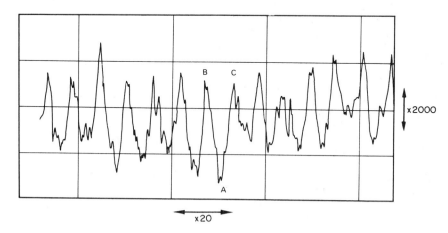

Fig. 4.41. Trace of surface illustrating distortion due to different magnifications.

THE TOMLINSON SURFACE METER

This instrument, shown in fig. **4.42**, uses mechano-optical magnification methods and was designed by Dr. Tomlinson of the National Physical Laboratory.

The skid is attached to the body of the instrument, its height being adjustable to enable the diamond-tipped stylus shown to be positioned conveniently. The stylus is restrained from all motion except a vertical one by a leaf spring and a coil spring; the tension in the coil spring P causes a corresponding tension in the leaf spring. These forces hold a cross-roller in position between the stylus and a pair of parallel fixed-rollers as shown in the plan view. Attached to the cross-roller is a light-spring steel arm carrying at its tip a diamond which bears against a smoked-glass screen.

In operation, the body of the instrument is drawn slowly across the surface by a screw turned at 1 rev/min by a synchronous motor, the glass remaining stationery. Irregularities in the surface cause vertical movements of the stylus which cause the cross-roller to pivot about point A and thus produce a magnified motion at the marking diamond on the arm. This motion, coupled with the horizontal movement, produces a trace on the glass magnified in the vertical direction at 100×, there being no horizontal magnification.

The smoked-glass is transferred to an optical projector and magnified further at 50×, giving an overall vertical magnification of 5000× and a horizontal magnification of 50×. The trace may be taken off, by hand or by photographic methods, and analysed.

THE RANK TAYLOR-HOBSON 'TALYSURF'

The Talysurf is an electronic instrument which

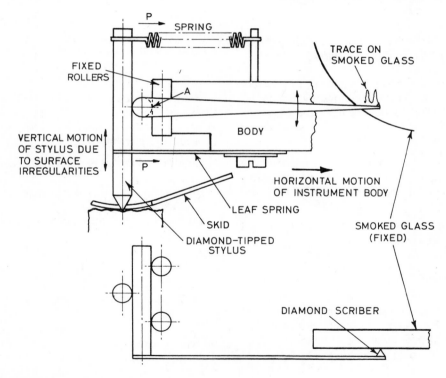

Fig. 4.42. Tomlinson Surface Meter.

works on the same basic principles as the Tomlinson instrument, but the methods of magnification and recording the trace are different. It gives the same information much more rapidly and probably more accurately. It can be, and is, used on the factory floor or in the laboratory, whereas the Tomlinson instrument is essentially for use in the laboratory.

The measuring head again consists of a stylus, and a shoe which controls the movement of the instrument head across the surface. In this case the stylus movements relative to the shoe produce an electrical signal which can be used to produce a trace and a numerical assessment, the whole operation taking about two minutes. With some Talysurf instruments, no recorder is provided and the instrument simply gives a numerical assessment directly on a dial.

NUMERICAL ASSESSMENT OF SURFACE ROUGHNESS

Specific Objective: *The student should understand the meaning of symbols R_a and R_z.*

The numerical value applied to surface roughness which is most commonly used in this country is the *Arithmetical Mean Deviation* denoted by the symbol R_a. The arithmetical mean deviation is defined as:

the arithmetical average of the departure of the profile above and below the reference line (centre or electrical mean line) throughout the prescribed sampling length.

Thus, referring to fig. 4.43, if equally spaced ordinates are erected at $1, 2, 3, 4, \ldots n$, whose heights are $h_1, h_2, h_3, h_4 \ldots h_n$, then,

$$R_a = [h_1 + h_2 + h_3 + h_4 + \ldots h_n]/n$$

Note that in this expression for R_a, the heights of the ordinates must be entered regardless of sign. If the weight of each ordinate was measured from a mean line with regard to sign their average value would always be 0.

To determine an R_a value from a trace by the erection of ordinates would be a laborious

process, and if an unfortunate ordinate spacing was chosen, significant points on the surface might not be included. However, if an irregular area is divided by its length, the value obtained is the average height of the area. Such an area can be measured readily by a planimeter, and if the total area enclosed by the trace and the mean line is divided by the length of the trace, the average height of the trace from the mean line is obtained. Thus, referring to fig. 4.44,

$$\text{Average height of trace} = \frac{A_1 + A_2 + A_3 + \ldots A_n}{L}$$

This value obtained is for the trace, not the surface under test. To obtain the R_a value for the surface, the value must be divided by the vertical magnification and multiplied by 10^3 to give the R_a value.

$$R_a \text{ value} = \frac{A_1 + A_2 + A_3 + \ldots A_n}{L} \times \frac{10^3}{\text{vertical magnification}}$$

where the sum of the values of
A = total area in mm^2;
L = length of trace in mm.

Before the area is measured, the position of the mean line must be fixed. This can again be done by using a planimeter, by the following procedure. Referring to fig. 4.45,

(a) draw a line AB parallel to the general line of the trace;
(b) enclose the area by parallel end lines AA_1 and BB_1;
(c) measure the total area by planimeter. Then,

$$h = \frac{\text{total area}}{\text{length}}$$

where h = the distance of the mean line from AB.

This technique to analyse the trace is required for the Tomlinson instrument. The Talysurf incorporates an integrating device which gives the result directly on an average meter.

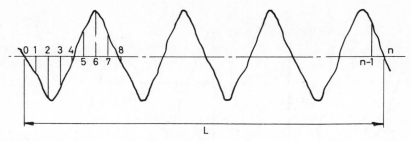

Fig. 4.43. Determination of R_a value by measuring ordinates.

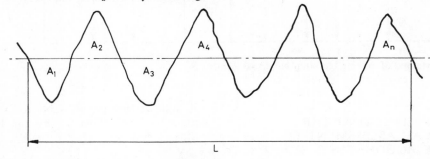

Fig. 4.44. Determination of R_a value by measuring areas.

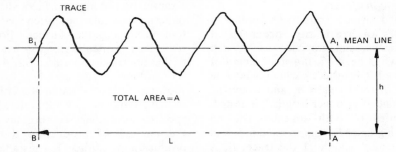

Fig. 4.45. Fixing the position of the mean line of a surface trace. $h = \dfrac{\text{TOTAL AREA}}{L}$

The ten point height of irregularities (R_z)

In some cases it is more convenient to have an assessment of peak-to-valley height of the surface irregularities. This is obtained by drawing a reference line A_1B_1 parallel to the mean line AB drawn through the trace. The heights of the five highest peaks and the five deepest valleys are then measured from line A_1B_1, as shown in fig. 4.46. If these heights h_1, h_2, h_3 are measured in mm from a trace, then,

$$R_z = \frac{(h_2 + h_4 + h_6 + h_8 + h_{10})}{5}$$
$$-\frac{(h_1 + h_3 + h_5 + h_7 + h_9)}{5}$$
$$\times \frac{10^3 \, \mu m}{\text{vertical magnification}}$$

115

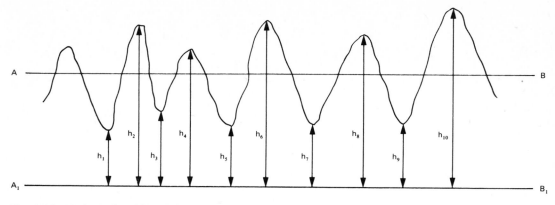

Fig. 4.46. Method of making ten point height assessment R_z.

FACTORS EFFECTING THE NUMERICAL ASSESSMENT OF SURFACE ROUGHNESS

1. *Effect of length of trace*

Most machined surfaces consist of a number of different types of irregularity, occurring at different spacings, superimposed upon each other as shown in fig. 4.47. If a short length of surface of length l_1 is tested, it will only include irregularities of total height h_1 and a low R_a value will result. If a longer length l_2 is tested, all irregularities of both spacings will be included and a greater R_a value related to h_2 will result. If the total length of the trace, l_3, is considered in the determination of R_a, the result will be much greater and related to h_3.

It follows that the R_a value for a complex surface will be influenced by the length of surface sampled.

2. *Effect of lay of surface*

Examination of most surfaces shows a regular pattern of irregularities, often feed marks of the tool producing the surface. For functional reasons, it is often necessary that the direction of these marks be specified. B.S. 1134 provides six symbols specifying directions of lay as shown in fig. 4.48. Examination of the meaning of these symbols shows that, to a great extent, specification of the direction of lay also specifies the method of production to be used in producing the surface. For instance it would be extremely difficult to produce surfaces to symbols 'C' and 'R' on a shaping machine.

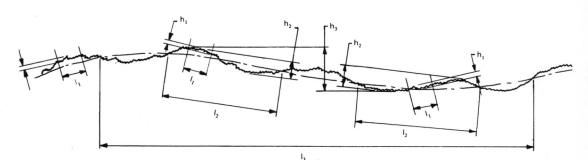

Fig. 4.47. Effect of sample length of surface on numerical assessment.

Symbol	Interpretation	
=	Parallel to the plane of projection of the view in which the symbol is used. 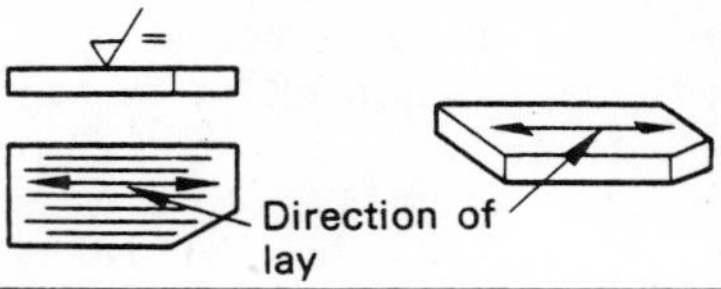	
⊥	Perpendicular to the plane of projection of the view in which the symbol is used. 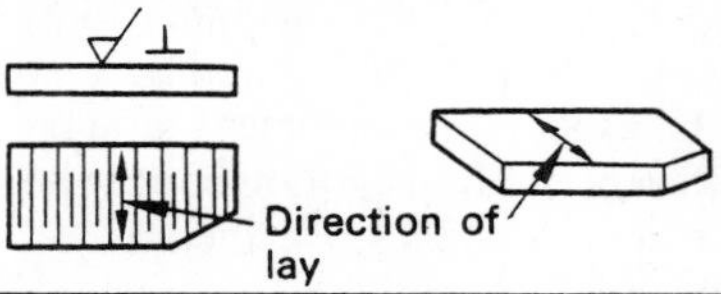	
X	Crossed in two slant directions relative to the plane of projection of the view in which the symbol is used. 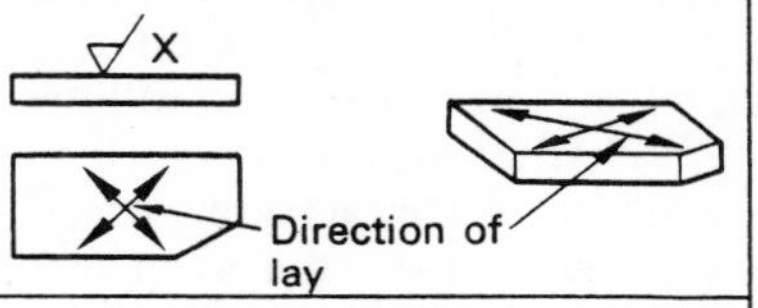	
M	Multi-directional. 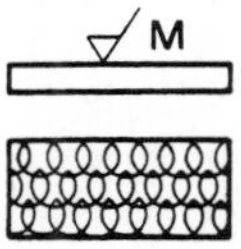	
C	Approximately circular relative to the centre of the surface to which the symbol is applied. 	
R	Approximately radial relative to the centre of the surface to which the symbol is applied. 	

NOTE 1. Should it be necessary to specify a direction of lay not clearly defined by these symbols, this may be done by a suitable note on the drawing.

NOTE 2. See BS 308, Part 2 (in course of preparation), Appendix B, for method of indication on drawings.

Fig. 4.48. Symbols for the direction of lay (from: BS 1134: Part 1: 1972). Reproduced by permission of the British Standards Institution.

It must also be realised that in measuring surface texture, it is necessary to determine the maximum R_a value for the surface. Consider a surface consisting of regular vee-grooves along the length of the surface, the sides of the groove being perfectly smooth. If the roughness is measured at 90° to the direction of the grooves (across the lay) a maximum R_a value will be obtained. If the roughness is measured along the grooves, the R_a value will be zero, measurements in directions between these two resulting in intermediate R_a values.

The roughness should always be measured in a direction to give the maximum R_a value which, except in the case of surfaces with a multi-directional scratch pattern, will always be at right-angles to the direction of lay of the surface.

3. *Effect of type of surface*

Although the lay of the surface specified does, to some extent, decide the method of manufacture to be used, some surfaces could be produced by different methods and produce the same direction of lay. Surfaces specified as '=' could be produced by shaping or surface grinding. Now consider the surfaces depicted in fig. 4.49(a) and (b). These surfaces will be totally different in character, a surface such as (b) having a small initial contact area which increases slowly as wear takes place; consequently the surface would wear very rapidly. Conversely, the surface (a) has a contact area which increases very rapidly with wear and the rate of wear will be much less. Examination of these surfaces shows them to be of identical form, one being the mirror image of the other, and will have the same R_a values, but they have opposite characteristics. It follows that all these factors must be considered in specifying and measuring a surface and that R_a value alone does not give enough information to enable surface qualities to be sensibly compared. Equally the trace is valueless unless its magnification in both horizontal and vertical directions is stated. Thus the information

necessary for completely specifying a surface is as follows:

1. R_a (or R_z) value.
2. Direction of lay.
3. Method of manufacture.

When the surface texture is measured the following factors must be considered and information given:

1. Direction of lay – measurements should always be made at right angles to the direction of lay to obtain the maximum R_a value for the surface.
2. Method of machining – values should only be compared for surfaces which have been produced by the same method of manufacture.
3. Meter cut-off – R_a values should only be compared if obtained from the same sample length of surface or meter cut-off value.
4. Trace – sensible comparisons can only be made between traces, having the same horizontal and vertical magnifications. If this is not possible, the ratio between the two magnifications should be the same e.g., 5000× vertical and 50× horizontal may be compared with 2000× vertical and 20× horizontal magnifications.

QUALITIES OF SURFACE FINISH

The quality of surface produced by a process depends on many features. A turned surface, for instance, will be rougher if the feed is coarse than if a fine feed is used. A very coarse grit

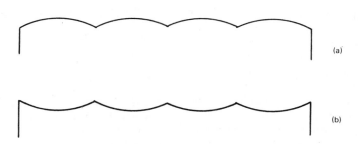

Fig. 4.49. Two surface textures have identical R_a values but opposite mechanical characteristics.

wheel used for rough grinding will produce a rougher surface than a wheel used for finish grinding, which may indeed produce a finer surface finish than coarse lapping.

The table shows the range of surface finish values which can be expected from various machining processes. Where a single figure is given, it is the best surface finish that can be expected from the process. When a surface texture is specified on a drawing, a single value is usually given, not the limits between which the finish produced must lie. In this case the finish stated is the WORST that may be allowed.

reconditioned, it is necessary that these alignments be tested before the machine is accepted and it is often the lot of the technician to carry out these tests.

The principles of the tests are similar for most machine tools and it is these principles which will be explained here rather than the detailed tests for individual machines.[2]

In the manufacture of machine tools, it is most important that slideways are straight and parallel to one another. Straightness tests are normally carried out using a microptic auto-collimator and the technique will only be

Process	Expected surface finish ($\mu m R_a$)
Rough turning	6.3
Rough grinding	3.2
Shaping and planing	1.6
Milling (H.S.S. tools)	0.8
Drilling	0.8–12.5
Finish turning	0.4–6.3
Reaming	0.4–3.2
Commercial grinding	0.4–3.2
Finish milling	0.2–1.6 (using negative rake cutters)
Diamond turning	0.1–0.8
Finish grinding	0.05–0.4 (precision work)
Honing and lapping	0.025–1.6

ALIGNMENT TESTING OF MACHINE TOOLS

Specific Objective: *The student should be able to use test mandrels and spirit levels for alignment testing of machine tools.*

In previous books covering TEC units at levels 1 and 2[1] in this subject, the author has outlined the alignments of various machine tools which must be correct if the machine is to produce work of the correct geometric form. If a new machine is purchased or an old one has been

discussed in specialist works dealing with metrology.[3] However, twist or wind in slideways can be induced during installation, and tests for this and other misalignments are carried out using a sensitive spirit level. Other tests are carried out using test mandrels and dial gauges to ensure that, after installation, the machine alignments are correct.

[2] Specific tests for machines of all types are given in the book *Testing of Machine Tools* by Dr. G. Schlesinger and in *Acceptance Test Charts for Machine Tools* prepared jointly by the Institution of Mechanical Engineers and the Institution of Production Engineers.

[3] *Metrology for Engineers* by Galyer and Shotbolt (Cassell) discusses the use of the microptic auto-collimator.

[1] *Technician Workshop Processes and Materials 1*, (Cassell) and *Technician Manufacturing Technology 2*, (Cassell).

THE PRECISION LEVEL AND ITS APPLICATIONS

A spirit level is simply a glass tube, or vial, curved to a radius R, and partially filled with liquid, leaving a bubble which always rises to the highest point of the curve.

If the vial is mounted in a base of length L, and one end of the base is raised through a height h then the bubble, moving to the highest position, will move a distance x along the tube as shown in fig. 4.50. Due to the change in height of one end of the base, the level is tilted through an angle δ radians, and:

$$\delta = \frac{h}{L} \text{ radians}$$

Recalling that,

$$\text{angle in radians} = \frac{\text{arc length}}{\text{radius}}$$

then

$$\delta = \frac{x}{R} = \frac{h}{L}$$

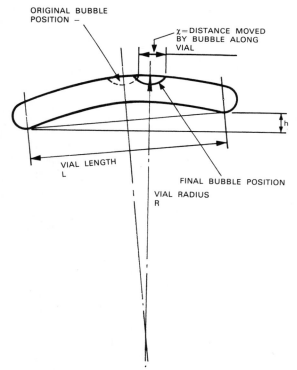

ORIGINAL BUBBLE POSITION –

x = DISTANCE MOVED BY BUBBLE ALONG VIAL

VIAL LENGTH L

FINAL BUBBLE POSITION

VIAL RADIUS R

h

Fig. 4.50. Principal of spirit level.

Considering the spirit level as a high magnification measuring device it is seen that:

$$\text{magnification} = \frac{\text{bubble movement}}{\text{height change of base end}}$$
$$= \frac{x}{h} = \frac{R}{L}$$

Thus for high magnification a large radius R of the vial is necessary. Consider a spirit level having a graduated vial, with the graduations marked at 2.5 mm intervals, mounted in a base 250 mm long, each graduation to represent a change in height of 0.005 mm of one end of the base.

$$R = \frac{x \times L}{h} = \frac{2.5 \times 250}{0.005} \text{ mm}$$
$$= 125\,000 \text{ mm}$$
$$R = 125 \text{ m}$$

The vial may be mounted in a straight base or in an accurately machined square block in which case it is known as a block level.

Testing for level and wind

When a machine is installed it should be level and, more important, the bed should not be twisted, thus inducing wind or non-parallelism in the slideways. In the case of a lathe, the level is tested as shown in fig. 4.51, the level being placed at a–a, to give the level in the longitudinal direction. Twist or wind is tested by placing the level across the slideways at each end of the bed, on a bridge-piece if necessary. In the case of vee/flat slideways, a special bridge-piece as shown in fig. 4.52 is used.

Testing for squareness

The block level is conveniently used for testing the squareness of vertical slides with horizontal slides and tables as shown in fig. 4.53. The level is first placed on the table and the reading noted. It is then placed against the vertical slideway, the difference in readings being the error in squareness, expressed either as an angle or in µm per unit length of the base of the block level.

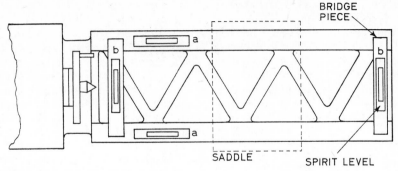

Fig. 4.51. Testing a lathe bed for level and wind.

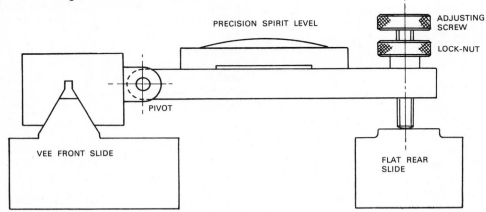

Fig. 4.52. Testing machine slides for wind or twist.

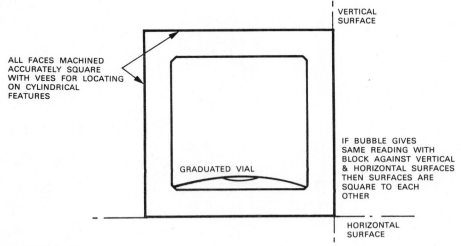

Fig. 4.53. Block level.

121

APPLICATIONS OF TEST MANDRELS AND DIAL GAUGES

Many alignment tests are carried out using test mandrels and dial gauges, the mandrels being hardened and ground to a high degree of precision. They must be carefully stored and handled so that they do not become damaged and indicate errors in the mandrel rather than the machine. The major alignments to be tested are that:

1. Machine spindle-nose taper axis is concentric with the axis of rotation (taper runs true).

2. Machine spindle axis is parallel to the tool or work motion.

3. Machine spindle axis is normal to the table surface.

4. Table surface is parallel to its own motion.

1. *Running accuracy of the spindle-nose taper*

This test is carried out by inserting an accurately-made hardened and ground test bar in the machine spindle nose and testing it with a dial gauge at positions I and II, fig. 4.54, with the spindle slowly rotating. This test applies to all rotating spindle machines.

2. *Machine spindle parallel to tool or work motion*

The same test bar can be used for this test. The spindle remains stationary and the machine saddle or table is moved. The test is carried out in two planes (a) and (b) as shown in fig. 4.54. This test applies to lathe saddles, milling machine cross-slide table motions, vertical milling machine vertical slide motions, horizontal boring machines, to name but a few.

3. *Machine spindle axis normal to work table*

A test bar is used to support a dial gauge at a suitable radius of rotation so that the plunger bears on the table surface as shown in fig. 4.55. As the spindle is slowly rotated, variations in dial-gauge readings indicate non-squareness of the table to the axis of rotation. Note that this test is another application of the reversal principle discussed previously. The test applies to drilling machines, vertical-spindle milling machines, jig boring machines, etc.

4. *Table surface parallel to its own motion*

A dial gauge is mounted on a fixed member of the machine, with the plunger bearing on the machine table and the table is slowly traversed

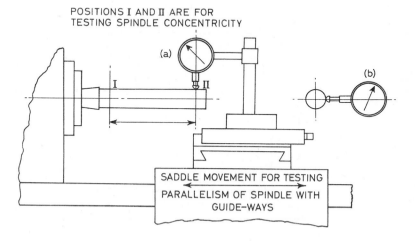

POSITIONS I AND II ARE FOR TESTING SPINDLE CONCENTRICITY

(a)

(b)

SADDLE MOVEMENT FOR TESTING PARALLELISM OF SPINDLE WITH GUIDE-WAYS

Fig. 4.54. Tests for spindle concentricity and alignment with saddle motion.

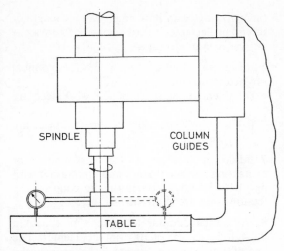

Fig. 4.55. Testing a drilling-machine spindle for squareness with the table.

under the gauge as shown in fig. 4.56. If the table surface is not parallel with its own motion, it produces a change in dial-gauge reading and, more important, non-parallel work. This test applies to milling machines, surface grinding machines, jig boring machines, horizontal boring machines, and shaping machine tables.

5. *Machine table tee-slots*

Table tee-slots, which are used to locate many fixtures in correct alignment with machine spindles, can be tested for these alignments using similar methods to (3) and (4) above.

A particular test which is worthy of mention is to check that the motion of a lathe cross-slide is at right-angles to the axis of rotation. This is best checked by setting up a large billet and facing across it to clean up the surface. The machined surface is then checked for flatness using a straight edge. If it is flat, then the tool motion is square to the axis of rotation of the work; if it is conical in any way then an error exists in the cross-slide motion.

The reason for particularly mentioning this test is that some people, believing that a dial gauge is the only reliable method of test, have been known to set a dial gauge up in the cross-slide and run it across the machined surface. As the dial gauge is moving in the same path as the tool which produced the surface no error will be detected however large it may be.

This also applies to a machine table which has been machined on the machine of which it was a part. Consider a magnetic chuck which has been cleaned-up on the surface grinder on which it is mounted. Its shape will conform to the table motions produced by the slideway geometry. Any errors in the slideways will be reproduced on the magnetic chuck, and will not be detected by mounting a dial-gauge to bear against the chuck and traversing the table. Flatness of the magnetic chuck must be checked independently of the table motion.

Never check an alignment by using a dial gauge on a surface machined in the machine under test.

QUESTIONS

1. A workpiece consists of a block of tool steel, hardened and ground, of overall height 245 $\pm$0.005 mm. The block is 100 mm square and has a groove ground across it, the height from the base of the block to the bottom of the groove being specified as 225 $\pm$0.008 mm.

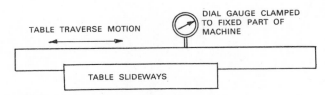

Fig. 4.56. Testing table surface for parallelism with table motion.

(a) List the equipment to check the height from the base to the bottom of the groove.

(b) Explain how the measurement would be carried out.

2. Electrical and pneumatic comparators have certain features in common which give them advantages over mechanical and optical comparators. List these features and explain what advantages they confer.

3. A measuring fixture is required to continuously check the thickness of rubber strips. Due to the softness of the rubber it distorts under normal measuring pressure and it cannot be guaranteed that the measuring area is clean or that the rubber is accurately positioned.

Sketch a suitable measuring head for the fixture and explain how its features overcome the above problems.

4. An optical projector is required and, due to the limitations of the room, the maximum distance from the lens to the screen is 5 metres although it can be less. Lenses are available having focal lengths of 100, 75 and 50 mm.

(i) Which lens should be used?

(ii) With the lens selected what should be the distance from the lens to the screen.

5. A four start thread of 2 mm pitch has a mean diameter of 20 mm. Calculate the helix angle.

The form of the thread is to be checked using a projector in which helix interference is overcome by turning the thread axis through the helix angle. By how much will the length of thread projected be foreshortened? Express the answer as a percentage.

6. Make a diagram of a modern optical projector showing clearly how:

(a) linear dimensions of the workpiece are measured;

(b) angular dimensions of the workpiece are measured.

7. Make a diagram of an instrument for measuring surface texture and compare it with fig. 4.15, labelling clearly on the diagram the features that are shown on fig. 4.15.

8. An optical comparator of the type shown in fig. 4.36 has a lens of 100 mm focal length. The line of action of the plunger is 5 mm from the mirror pivot and the eyepiece magnification is $30\times$. Calculate the magnification of the comparator.

9. A trace measured from a machined surface is shown in fig. 4.57. The trace has been measured and the areas and length of trace are shown, as are the magnifications used. Calculate the R_a value for the surface in micrometres.

10. An Angle Dekkor is set to combination angle gauges and a clear image is obtained. When a reading is taken on the workpiece it is readable but out of focus. Suggest a reason for this.

11. A hardened and ground wedge-shaped

AREAS IN CM² UNITS

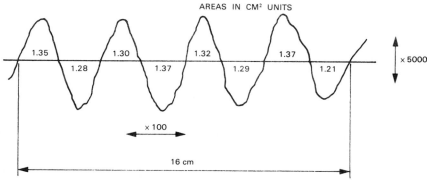

1.35   1.30   1.32   1.37
1.28   1.37   1.29   1.21

× 5000

× 100

16 cm

Fig. 4.57. Question 9.

block of steel has the angle between its two longest faces specified as 14° 32′ ±10′. The angle is to be checked using an Angle Dekkor.

   (i) List the equipment to be used.

  (ii) Write a detailed set of instructions for the inspector who is to carry out the measurement. Illustrate the instructions with diagrams where necessary, including views of the instrument eyepiece at various stages of the measurement.

12. The surface finish on the flat-faced surface of a turned workpiece is not to exceed 5 μm R_a, the measurement to be made at a meter cut-off of 0.8 mm and to be made at right angles to the machining marks (radially). Explain the meaning of this specification and how it would be shown on a drawing.

13. Explain with the aid of diagrams how the following alignment tests on machine tools would be carried out:

  (a) Spindle nose taper of a milling machine concentric with axis of rotation.

  (b) Axis of milling machine spindle parallel with table surface.

  (c) Axis of milling machine spindle parallel with table motion.

14. A technician is asked to check the alignments of a lathe and proceeds to machine a bar in the chuck. He then mounts a dial gauge in the saddle and runs it along the machined bar. This is followed by facing a block of metal in the chuck and checking it for flatness using a straight-edge. One test gives useful information and the other none.

  (i) State which test is wrong.

  (ii) Explain how it should have been correctly carried out.

ANSWERS

4. (i) 75 mm; (ii) 3.825 m.

5. 7.26°; 0.8%.

8. 1200×.

9. $R_a = 1.31$ μm.

5 Metal cutting with single point tools

INTRODUCTION

As the student has progressed through the level 1 and 2 units[1] of this subject he should have progressed from recognising the need for rake and clearance angles and the ability to recognise these angles, at level 1, to an ability to recognise the principles of metal cutting at level 2, where a simple analysis of the forces acting, and those forces which can readily be measured during orthogonal cutting, was made.

At the same time the student should now have some experience of using most of the machines in use in a general machine shop and, from his studies have some knowledge of the alignments which must be correct if the machines are to produce work of the correct geometric form.

Thus the background work has been completed and the stage has been reached where it is appropriate to analyse further the metal cutting process in the more practical case of oblique cutting rather than the somewhat unusual and simplified form of orthogonal cutting. Most of this chapter is therefore concerned with the metal cutting process applied to different types of cutter rather than with machines.

General Objective: *The student should understand the principles of metal cutting using a single point tool.*

In order to understand the principles of metal cutting, it is necessary to consider the different variables which together influence the cutting process, the relationship between these variables and the factors which enable an optimum condition to be arrived at for a given operation. This work is most easily carried out in practice using a single point tool under continuous cutting conditions, that is, in a lathe. The principles thus demonstrated can then be applied to multi-point interrupted cut cases with suitable modifications. The main reason for this is that it is relatively easy to measure forces on a stationary single point tool. It is not too difficult to measure the torque and thrust force involved in drilling, again a continuous cut, but the student may like to consider how to measure the forces on, or applied by, a helical tooth-slab milling cutter. It can be done, with difficulty, but it is outside the scope of this work. However it can be shown that the forces are similar to those in oblique single point cutting – they are simply applied differently. The variables which will be considered in this work are:

1. Forces acting during single point cutting.
2. Power consumed.
3. Metal removal rates.
4. Cutting speed and tool life.

Specific Objective: *The student should be able to name the three principal components of cutting force, resolve them into the cutting force acting and calculate power consumption and metal removal rates.*

---

[1] *Technician Workshop Processes and Materials 1* (Cassell) and *Technician Manufacturing Technology 2* (Cassell).

FORCES ACTING DURING METAL CUTTING

In level 2 it was shown that during an oblique metal cutting process it is convenient to measure three forces acting on a lathe tool. These forces are shown in fig. 5.1, and described below, and are seen to be mutually at right angles to each other.

1. TANGENTIAL FORCE (F_t)

As shown in the diagram this force is tangential to the work face and is in the direction of motion of the work relative to the tool. It is therefore the main power consuming force being the largest of the three and associated with the highest speed.

2. RADIAL FORCE (F_r)

This force acts normally to the work surface and would force the tool out of the work were their relative positions not fixed. As there is no motion in this direction this force consumes no power.

3. AXIAL FORCE (F_a)

The axial force is the one which resists the feed motion of the tool parallel to the axis of the work. As there is a motion associated with this force, power is consumed, but the feed motion is usually so slow that the amount of power consumed is, relatively, so small compared with that due to F_t that it can be ignored.

These forces are measured during cutting using a *cutting tool dynamometer*, as shown in fig. 5.2(a). Most dynamometers work on the principle that if a force is applied to an elastic member it deflects and the magnitude of the deflection is a measure of the force applied. The tool is carried, as shown, in a holder mounted on the front of a disc or diaphragm which is clamped to the main body of the dynamometer. The back of the diaphragm supports a bar against which dial-gauges bear to detect the movements of the bar caused by the tool forces deflecting the diaphragm. A third dial-gauge (not shown), operating in the horizontal plane, can be used to measure the third force, F_a, if necessary.

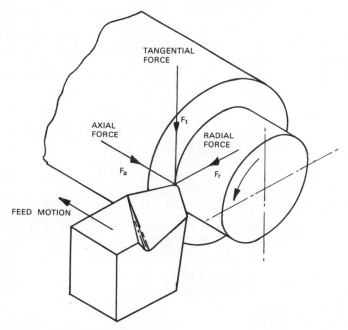

Fig. 5.1. Oblique cutting – three force system.

The tool is first replaced by an adaptor carrying a ball which is then loaded by known dead weights and a graph or calibration chart is made of load against deflection as shown in fig. 5.2(b). In operation, the tool should project a fixed amount so that its point is in the position of the ball centre. Its deflections are noted during cutting and by referring these deflections back to the calibration chart, the forces involved can be determined.

Knowing these forces, which are mutually at right angles to each other, their resultant, 'R', the force which is actually causing cutting to take place, can be determined. Assume that during a cutting operation the dial gauge readings obtained were 0.050 mm for F_r, 0.060 mm for F_a and 0.80 mm for F_t. From the calibration chart it can be seen that:

$$F_r = 300 \text{ kN} \qquad F_a = 630 \text{ kN} \qquad F_t = 950 \text{ kN}$$

These may be drawn as vectors as shown in fig. 5.3 and the resultant force R can be seen to be the resultant of forces R_1 and F_t where R_1 is itself the resultant of the two forces F_a and F_r in the horizontal plane. Using the theorem of Pythagorus:

$$R_1^2 = F_a^2 + F_r^2$$
$$R_1 = \sqrt{[(630)^2 + (300)^2]}(\text{kN})$$
$$R_1 = 697.78 \text{ kN}$$
$$R^2 = R_1^2 + F_t^2$$
$$R = \sqrt{[(697.78)^2 + (950)^2]}$$
$$= 1178.72 \text{ kN}$$

To find the direction in which R acts relative to the work, its is necessary to find the angles θ_1 and θ_2 in fig. 5.3. and express the direction as θ_2° to the horizontal in a vertical plane at θ_1 to the work axis. From fig. 5.3:

$$\text{Tan } \theta_1 = \frac{300}{630}$$
$$\theta_1 = 25.46^\circ$$
$$\text{Tan } \theta_2 = \frac{697.78}{950}$$
$$\theta_2 = 53.70^\circ$$

Thus the direction of the force acting on the tool relative to the work is *53.70° to the horizontal in a vertical plane aligned at 25.46° to the work axis*. In fact this is rather difficult to visualise and it is always better to draw a diagram of the type shown in fig. 5.3.

If a little thought is given to the matter, it will be seen that for the process to remain in equilibrium, an equal and opposite force is applied by the tool to the work and it is this force which causes the chip to shear off the parent metal and slide over the tool face.

It was shown in the previous work[1] that the chip shears ahead of the tool and, in fact, the shear area remains relatively constant as long as the feed and depth of cut remain unchanged.

[1] *Technician Manufacturing Technology 2* (Cassell).

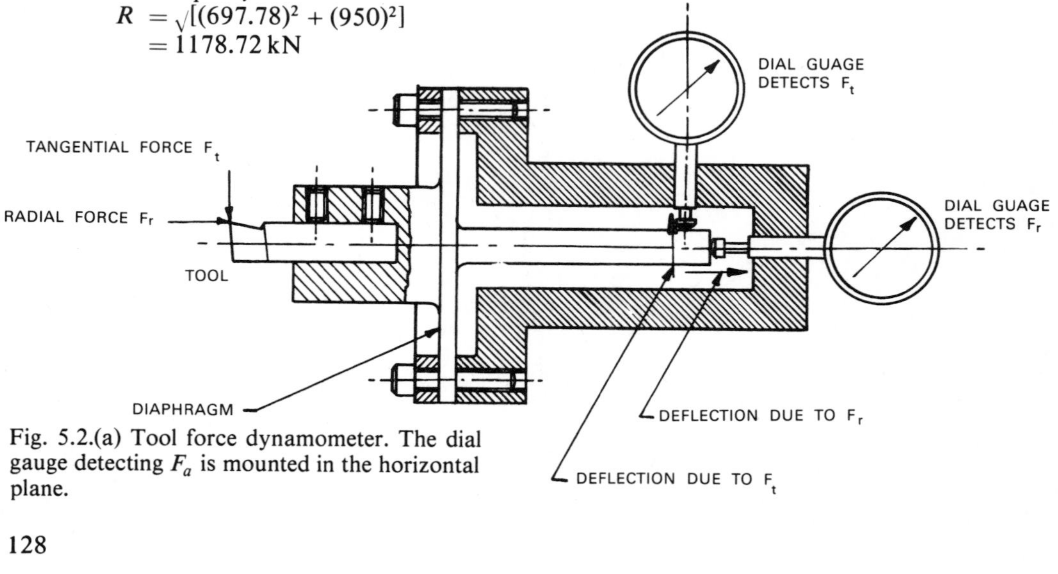

TANGENTIAL FORCE F_t

RADIAL FORCE F_r

TOOL

DIAPHRAGM

DIAL GUAGE DETECTS F_t

DIAL GUAGE DETECTS F_r

DEFLECTION DUE TO F_r

DEFLECTION DUE TO F_t

Fig. 5.2.(a) Tool force dynamometer. The dial gauge detecting F_a is mounted in the horizontal plane.

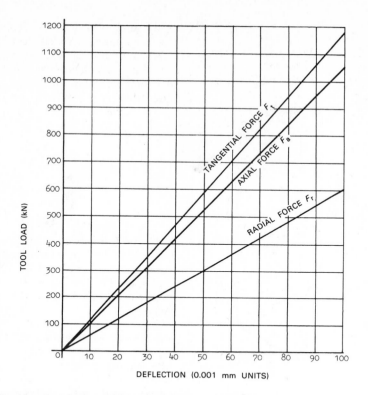

Fig. 5.2.(b) Calibration chart for use with cutting tool dynamometer.

Thus if the cutting speed is increased, the depth of cut and feed per revolution remaining constant, the cutting force should not change. However as:

$$\text{Power} = \text{Force } (N) \times \text{velocity } (m/s)$$

it is to be expected that the power required will increase with the cutting speed. It can be shown by experiment that this is very close to what actually happens. In fact the cutting force is high at very low speeds and as the cutting speed, i.e., peripheral speed (in metres/min) increases, the force drops to a minimum and then rises very slowly as the speed increases. The table and graph in fig. 5.4, show this for a cutting operation on a lathe using a high-speed steel tool having the conditions as follows:

clearance angle 6°;
true rake 20°;
plan approach angle 30°;
depth of cut 10 mm;
feed per rev. 0.12 mm.

POWER CONSUMED DURING METAL CUTTING

The power consumed during a cutting operation is often measured using a watt meter which records the power supplied to the machine motor. If the machine is run at the speed at which the cut is to be made, with the

129

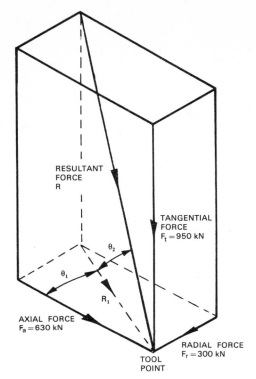

RESULTANT
FORCE
R

TANGENTIAL
FORCE
$F_t = 950$ kN

θ_2

θ_1

R_1

AXIAL FORCE
$F_a = 630$ kN

RADIAL FORCE
$F_r = 300$ kN

TOOL
POINT

Fig. 5.3. Resolution of three force system into the resultant force R, the force acting on the tool.

feed engaged and the watt meter reading noted, and then the reading noted during cutting, it is assumed that:

power due to cutting
= (power during cutting) − (no load power)

This is not a good way to determine the power absorbed in the cutting operation, because when the tool reaches its full depth of cut the cutting force is transmitted to the slideways and bearings of the machine and the power lost to friction is increased by an amount which remains unknown.

It is better to measure the tangential force, F_t, and determine the surface speed of the work knowing the diameter and rev/min. Alternatively, the surface speed can be determined directly using a tachometer fitted with a rubber disc, which is held against the

130

work surface as it rotates. The disc is made to a specific diameter so that the tachometer scales read the surface speed direct in metres/min.

Assume that a cut taken on a bar of 150 mm diameter at 200 rev/min produces a measured tangential force of 900 N on the work.

$$\text{Surface speed} = \pi \times \frac{150\ \text{mm}}{1000} \times \frac{200}{60}\text{rev/sec}$$
$$= 1.57\ \text{metres/sec}$$
$$\text{power (watts)} = \text{force (newtons)}$$
$$\times\ \text{velocity (metres/sec)}$$
$$= 900\ \text{N} \times 1.57\ \text{m/s}$$
$$= 1413\ \text{watts}$$
$$\text{or} = 1.413\ \text{kW}$$

Of itself, this is of no great interest since, as we know that the force remains relatively constant, the power will increase as the speed increases. The main interest is to compare the power consumed with the rate of metal removal.

RATE OF METAL REMOVAL

The simplest expression of metal removal rates is in terms of $(\text{mm})^3/\text{min}$. As the depth of cut is known and the feed rate is known then:

$$\text{Area of cut} = \frac{\pi}{4}\ (D_1^2 - D_2^2)$$

where D_1 = diameter before cutting;
and D_2 = diameter after cutting

If during each revolution of the work the tool advances a distance, f mm then:

volume removed/rev = area of cut × feed/rev

$$= \frac{\pi}{4}\ (D_1^2 - D_2^2)f\ \text{mm}^3/\text{rev}$$

where f = feed rate in mm/rev

As the work is rotating at N rev/min, then:

volume removed/min

$$= \frac{\pi}{4}\ (D_1^2 - D_2^2)fN\ \text{mm}^3/\text{min}$$

This again is of little direct interest; it tells us roughly how much swarf we are going to make, but if it is compared with the *power con-*

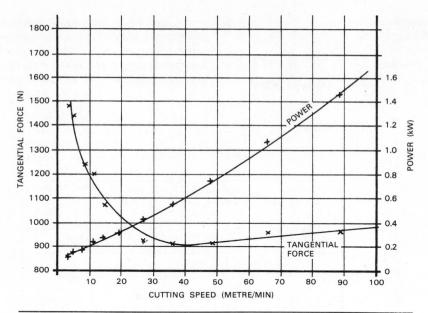

Fig. 5.4. Results of cutting test and graphs of cutting speed (m/min.) v tangential force (F_t) and power consumed (kW).

Cutting Speed (S) (m/min)	Tangential Force (F_t) (N)	Power (kW) $\dfrac{F_t \times S}{60 \times 1000}$
4.50	1480	0.111
6.00	1440	0.144
8.20	1240	0.170
11.20	1200	0.224
14.75	1080	0.266
19.90	960	0.318
27.00	910	0.410
36.00	910	0.546
49.00	910	0.743
66.70	960	1.056
89.70	960	1.435

sumption at the tool during the cutting process then we determine *how much metal can be removed per minute for the expenditure of 1 kW of power.*

Relating this to the previous examples, we find that for a depth of cut of 10 mm on a 150 mm bar at 200 rev/mm and 0.12 mm/rev feed rate:

$$\text{Area of cut} = \frac{\pi}{4}(150^2 - 130^2)$$
$$= 4398.23 \text{ mm}^2$$

Volume removed/rev
$$= 4398.23 \ (\text{mm}^2) \times 0.12 \ (\text{mm/rev})$$
$$= 527.79 \text{ mm}^3/\text{rev}$$

Volume removed/min
$$= 527.9 \ (\text{mm}^3/\text{rev}) \times 200 \ (\text{rev/min})$$
$$= 105\ 558 \text{ mm}^3/\text{min}$$

131

But this requires an expenditure of 1.319 kW at the tool and therefore:

$$\text{Volume/min. kW} = \frac{105\,588\ (\text{mm}^3/\text{min})}{1.413\ (\text{kW})}$$
$$= 74\,726\ \text{mm}^3/\text{min kW}$$

If the table associated with fig. 5.4 is completed to give this value for each cutting speed, a graph may be drawn of vol/min. kW against cutting speed as shown in fig. 5.5. It can be seen that this rises to a peak and then falls

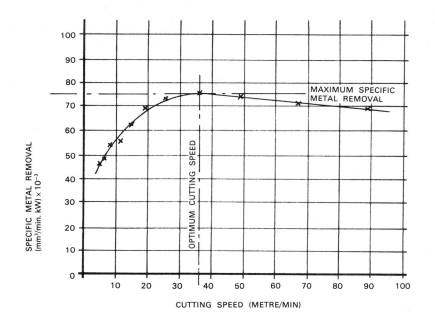

Cutting speed, S (m/min)	Tangential force, F_t (N)	Power (kW) $\dfrac{F_t \times S}{60 \times 1000}$	Vol/min mm³/min	Specific metal removal rate (mm³/min.kW)
4.50	1480	0.111	5277.9	45 894
6.00	1440	0.144	6891.3	46 673
8.20	1240	0.170	9500.2	54 286
11.20	1200	0.224	12 667	56 048
14.75	1080	0.266	16 889	62 321
19.90	960	0.318	22 167	69 927
27.00	910	0.410	30 084	73 916
36.00	910	0.546	40 112	74 975
49.00	910	0.743	54 890	73 876
66.70	960	1.056	74 946	70 042
89.70	960	1.453	100 280	69 979

Fig. 5.5. Results of cutting test and graph of specific rate of metal removal (mm³/min. kW) v cutting speed (m/min.).

132

off, the peak value being the *maximum* amount of material which can be removed for the expenditure of 1 kW power with this tool at this depth of cut and at this feed rate. This then should give the most economical cutting speed for these conditions as shown.

The cutting tool dynamometer has a further, more significant, use in research into metal cutting. It can not only show the effects of cutting speed, changes in feed rate and depth of cut on tool forces, but if a set of cutting conditions are fixed by the work to be done, it can be used to determine the best tool geometry in terms of rake angle and plan approach angle for new materials. Many years ago the author was privileged to see the results of such work carried out at Cranfield Institute of Technology on the then new Nimonic alloys being used for gas turbine blades. These blades were being machined by end milling and the cutter life was of the order of 20 min. By testing tools of different shapes in a systematic manner the group at Cranfield were able to quadruple the tool life at increased cutting speeds and metal removal rates. Thus the cutting tool dynamometer is an extremely powerful research tool particularly in the fields of cutting new materials.

CUTTING SPEED AND TOOL LIFE

Specific Objective: *The student should calculate tool life given the expression $VT^n = C$.*

Under modern cutting conditions a tool usually fails in one of two ways:

(a) A land is worn on the clearance face, thus destroying the clearance angle and giving rise to rubbing, as shown in fig. 5.6(a).

(b) A crater is worn in the rake face behind the tool edge.

If the crater is allowed to develop, the cutting condition changes to one of high positive rake, giving rise to weakness and a calamitous tool failure as shown in fig. 5.6(b). Either of these conditions can demand an excessive amount of

tool regrinding and in the case of cemented-carbide-tipped tools is an expensive business. A tool is therefore considered to have failed when either:

(a) the wear land exceeds 0.25 mm; or
(b) at the onset of cratering.

Modern production methods require that the tool life should be known and ideally last longer than one work shift, but a complete new set of pre-set tools is kept ready to be fitted into the machine with the minimum of difficulty. At the end of the shift the old tools are removed before they have failed, and are replaced by the set of new ones, for the next shift. This avoids replacing individual tools at irregular intervals, with a consequently greater machine 'down-time'. The tool life under the known cutting conditions can be found by testing under these conditions. It is fortunate that there is a mathematical relationship between the tool life T (in minutes) and the cutting speed, V (in metres per minute). The relationship is given by

$$VT^n = C$$

where n and C are constants depending on the conditions, i.e., depth of cut, feed rate, coolant, etc.

Because the tool life and cutting speed are related in this way the tests can be accelerated by running at much higher speeds than occur in practice. From the values obtained for n and C in these tests a value for V can be found to give an adequate tool life T.

It must be emphasised that the value of T is the *actual cutting time*, not the time the tool is in the machine. Thus a tool life of 100 min could well be equivalent to a 240 min work shift in a capstan lathe where each tool is only used for part of the work cycle.

Consider a tool whose required tool life is 100 min. Tests carried out under the cutting conditions show that with a cutting speed of 200 m/min the tool life is 25 min, and with a cutting speed of 270 m/min the tool life is 15 min. The cutting speed which will give the required tool life of 100 min can now be calculated.

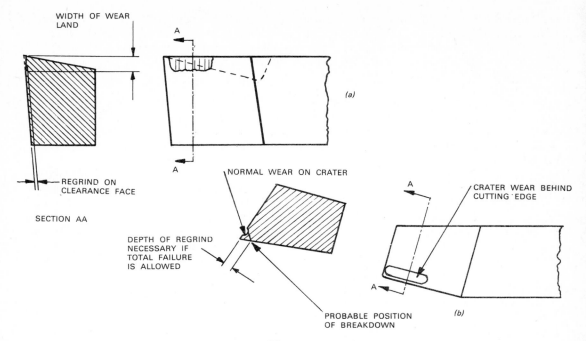

Fig. 5.6. (a) Wear land on flank face of tool, and (b) crater wear on rake face of cutting tool.

$VT^n = C$

$\therefore \log V + n \log T = \log C.$

Substitute known values,

$\log 270 + n \log 15 = \log C$

$\log 200 + n \log 25 = \log C$

$2.4314 + (n \times 1.1761) = \log C$ ...(1)

$2.3010 + (n \times 1.3979) = \log C$ ...(2)

Subtract

$0.1304 - (n \times 0.2218) = 0$

$\therefore n = \dfrac{0.1304}{0.2218}$

$n = 0.589$

Substitute in (1)

$2.4314 + (0.589 \times 1.1761) = \log C.$

$2.4314 + 0.689 = \log C$

$\therefore \log C = 3.1204$

$C = 1319.$

In practice the value of C is not required, since the values of n and $\log C$ are used to find the new value of the cutting speed V.

$\text{Log } V + n \log T = \log C$ $\log C = 3.1204$

                                    $T = 100 \text{ min}$

                                    $\log T = 2$

                                    $n = 0.589.$

Substitute:

$\text{Log } V + (0.589 \times 2) = 3.1204$

$\log V = 3.1204 - 1.178$

$= 1.9424$

$\therefore V = 87.6 \text{ m/min}$

An alternative solution can be used if we re-arrange the equation

$$\log V + n \log T = \log C$$

$$\text{to } \log V = -n \log T + \log C$$

in which n and $\log C$ are constants for a given set of conditions. This equation is therefore of the form $y = -mx + C$.

This is the equation to a straight-line graph with a negative slope.

Thus if we plot $\log V$ against $\log T$ for our known values, as in fig. 5.8, and draw a straight

134

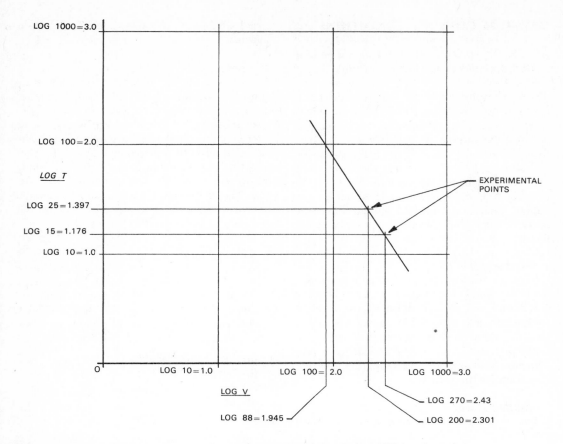

LOG 1000=3.0

LOG 100=2.0

LOG T

LOG 25=1.397

LOG 15=1.176

LOG 10=1.0

EXPERIMENTAL POINTS

O   LOG 10=1.0   LOG 100= 2.0   LOG 1000=3.0

LOG V

LOG 88=1.945

LOG 270=2.43

LOG 200=2.301

Fig. 5.7. Graph of log V and log T (V=cutting speed in m/min.; T=tool life in minutes).

line through them, the required value of log T=log 100=2 can be inserted and the corresponding value of log V=1.945 obtained. The required cutting speed is found from antilog V.

A word of warning on experimental method is important here. A straight-line graph can only be drawn through two experimental points if it is known that they conform to a straight-line law. In this case the work of many experimenters ensures us that this is so, but unless the law is known to operate these methods must not be used.

In practice it is unlikely that a speed of precisely 87.6 m/min could be obtained. The cutting speeds available depend upon the work diameter and the speed range available from the machine and the nearest *lower* speed should be used. This gives a further assurance that the tool life required will be exceeded.

It is not usually necessary to conduct tests on all the tools for a given machine set-up. Generally it can be decided from past experience which tool is working under the most arduous conditions and this tool is tested, the remainder working at suitable and convenient speeds.

OPTIMUM CUTTING CONDITIONS

Specific Objective: *The student should be able to state optimum conditions for cutting.*

The word optimum is derived from the Latin *optimus*, meaning best. The English definition which is most applicable to the above objective is 'most favourable or conducive to a given end'.[1] Thus it is required for a given cutting operation to determine the best conditions to be applied to achieve the desired objective. In metal cutting, objectives may differ, but for any cutting operation the tool should have the correct rake and clearance angles. Other factors to be considered are:

1. *Plan profile of tool*
Ideally the tool should have a plan approach angle and small nose radius to give good cutting conditions and surface finish. If, however, a square corner is required in the workpiece the student must consider whether it is worth sacrificing cutting efficiency and using a zero plan approach angle to save a separate squaring operation for the corner.

2. *Cutting speed*
It has been shown how the best cutting speed can be determined for cutting a given material with a given tool in terms of metal removal rates. However it may well be better to reduce the cutting speed to give a tool life which exceeds a shift length in order to avoid unnecessary tool changes.

Again the objective must be considered. In tool room or skilled work the best cutting speed should be chosen, life between regrinds being relatively unimportant. A special case is when a very long bar is being machined such as a roll for a rolling mill. Ideally a tool should not be changed during a cut and an ill-chosen speed could cause tool failure partway along a cut. In production work, when the tool lasts until a complete tool change is planned, the highest

[1] *Webster's Collegiate Dictionary* (G & C Merriam & Co.).

speed which will give this required tool life is selected.

3. *Feed rate*
The surface finish produced depends to a considerable extent on feed rate. Thus, if enough metal can be removed in one pass to give finished size, a fine enough feed must be chosen to give a suitable finish. If more than one cut is required then a roughing cut at a higher feed rate can be taken followed by a finishing cut to give a suitable surface finish at the final size. The feed rate for the roughing cut should be chosen not to overload the machine but certainly to make it work to its capacity.

4. *Cutting fluid*
Again the choice of cutting fluid depends upon the work being undertaken. High rates of metal removal can cause a built-up edge to form and extreme-pressure cutting oils are available which tend to prevent this. In general, then, the optimum conditions for cutting are:

(a) best cutting angles,
(b) best plan profile,
(c) best cutting speed,
(d) best feed rate,
(e) most suitable coolant,

to give the best rate of metal removal commensurate with tool life and surface finish to produce the workpiece in the most economic way. Each job must be considered on merit and its requirements taken into account when considering these factors.

TOOL CONSTRUCTION

Specific Objective: *The student should compare the advantages and limitations of various tool constructions including quick change tool holders and pre-set tooling methods.*

Since man first started to cut materials in power-driven machines, there has been a steady development in the type of cutting tool used. Initially the development was in cutting tool materials and, as has been shown in Book 2,[1] the need was for materials which retained their hardness at the high temperatures produced by the higher cutting speeds required. Although, as has been shown, higher speeds do not, of themselves, produce higher tool forces, the real need was for greater metal removal rates; if the optimum cutting speed was to be used this increase in metal removal rates could only be achieved by increasing the depths of cut and/or the feed rates. These increases do increase the tool force and so more rigid tools were required which would not deflect under these increased loads. Thus two types of single point tool were used:

1. *Tool bits made of high speed steel as small as 6 mm square where light cuts and low rates of metal removal were required.* These are still used on small machines under these conditions as the cost of the tool-bit is much less than that of the solid tool.

2. *Solid tools ground from larger pieces of H.S.S. mounted direct in the tool post where heavier cuts were to be taken.* Apart from lathes these were, and still are, also used in planing and shaping machines where heavy cuts are taken at relatively low cutting speeds. Cutting speeds were further increased with the development of cemented-carbide materials. These are too brittle to be used as a tool bit in a conventional tool holder or, indeed, as solid tools. In the early days of their use, cemented-carbide tools were brazed to a plain carbon steel shank and used either conventionally or with negative rake. Later still in the development of tool materials, came ceramic tool tips which could not be brazed to a shank but had to be clamped in place. Thus cutting tools had turned the full circle, almost back to the tool bit and holder of the early days but in the much more sophisticated form known as clamped-tip tools (*see* below).

CLAMPED-TIP TOOLS
If the tool to be used has a *negative* rake angle,

[1]*Technician Manufacturing Technology 2* (Cassell).

equal to the clearance angle, the angle between the rake and clearance faces is 90°. Thus a square slug of material, with a suitable nose radius on each corner, can be used directly as the cutting medium. Such a tool is shown in fig. 5.8. The top is clamped in position, well supported to resist the cutting forces and with a minimum amount of tip overhang. The shank is so designed that the tip is inclined about an axis along the cutting edge, this inclination providing rake and clearance, and the cutting edge is at constant height over the whole depth of cut. When the edge is worn, the tip is turned through 90° to bring a new edge into action. When all four top edges are worn, the tip is turned over to reveal four new cutting edges, giving a total of eight in all. When all the edges are worn the tip is disposed of, hence the name 'throw-away' tip.

Although the square-shaped negative rake tip has been used as an example, in modern developments, the tips need not be square, nor need they be for negative rake cutting only. Tips are made in the form of an equilateral triangle with a groove around the edge of the rake face, as shown in fig. 5.9 to give a positive rake and act as a chip breaker. The shape of tip and its cutting angle are chosen to be the most appropriate for the job in hand.

Neither is it necessary that the tips are clamped, as shown in fig. 5.8. Other types are made for other applications which are clamped by a central screw, this being more convenient for certain cases. In Chapter 6, the application of clamped throw-away tips to milling cutters is discussed, but their use is by no means limited to lathe tools and milling cutters. Fig. 5.10 shows how such tips may be applied to boring cutters.

Quick-change and pre-set tooling
A feature of all clamped-tip tools is the accurate location of the tip in the tool holder and the accuracy of manufacture of the tips themselves. Thus, when an edge is worn and the tip is indexed or replaced, the new cutting edge is accurately positioned to replace the old one and the work continues to be machined to the same

size as before. Thus the time taken to reset a production machine such as a capstan or turret lathe is considerably reduced.

In cases where indexable throw-away tips are not used, pre-set tooling may still be used to reduce the down-time involved in tool replacement. If the boring bars, knee-turning tools and roller boxes are accurately pre-set before they are fitted in the machine the old tools are removed and replaced with little or no setting adjustment needed before production can be resumed.

The actual tooling cost is higher because the situation is rather like that of the camper 'travelling light' – he needs three shirts, for example, one being worn, one ready to be washed and one clean for emergencies. In the same way with pre-set tooling, three sets are required – one in the machine, one being re-ground and re-set and one ready for use in a tooling rack at the side of the machine. This technique has been developed to a point where very large special-purpose machines such as transfer machines used for the complete machining of a motor vehicle cylinder block can be re-tooled in a very short time, usually in the period between shift-changes.

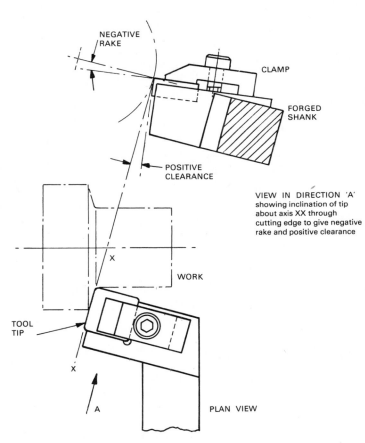

Fig. 5.8. Throw-away type clamped tip tool.

138

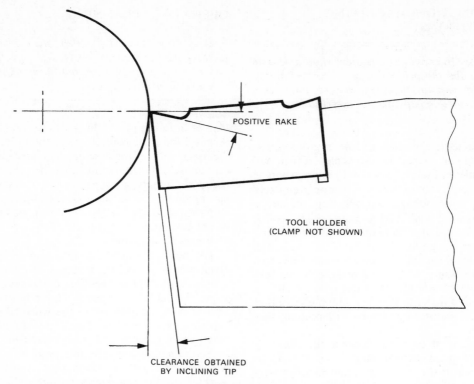

POSITIVE RAKE

TOOL HOLDER
(CLAMP NOT SHOWN)

CLEARANCE OBTAINED
BY INCLINING TIP

Fig. 5.9. Clamped tip tool with positive rake and chip breaker.

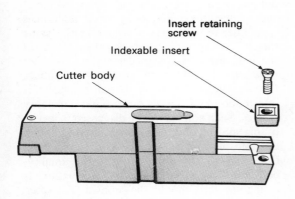

Insert retaining screw

Indexable insert

Cutter body

Fig. 5.10. Cutting head for boring bar showing method of locating and holding tips. Courtesy of Madison Tools Ltd.

Thus we return to the points considered at the beginning of this chapter which may be summarised as follows:

1. The tool should be chosen to give the most efficient and economic cutting possible under the conditions prevailing.

2. The cutting speed may not be the best possible but may be modified, using $VT^n = C$, to give the tool life required for the operation.

3. The construction of the tool should be chosen to suit the work in hand. Hand-ground solid tools and small tool bits may be used in jobbing shops, but modern production methods require quick-change tooling, either indexable clamped tool-bits or quick-change tools which can be pre-set and rapidly fitted in the machine with the minimum of setting adjustment.

PRODUCTION MACHINING OPERATIONS

General Objective: *The student should be able to describe the principles of process planning and the estimation of machining times and costs related to capstan and turret lathes.*

The discussion of cutting tools in this chapter, although of interest to the skilled turner and machinist in the toolroom and jobbing shop, is of particular concern to the planning engineer dealing with tooling for production operations on capstan and turret lathes. The same principles are applied in the planning of tooling and cam design for automatic lathes, although such machines are not considered at this level. It is necessary to be able to specify the tool shape which will give the most efficient cutting action and the cutting speed which gives either optimum cutting conditions or optimum tool life.

In addition to these factors, the planning engineer must ensure that the part will be produced in the most economic manner on a machine which is capable of doing the job. Finally he must be able to estimate the time required to produce the part and the cost of production. Thus, planning for capstan and turret lathes falls into the following stages:

1. Machine selection.
2. Operation layout.
3. Estimation of cutting time.
4. Cost estimation.

It can be seen that these procedures themselves fall into two categories, those concerned with the actual *operation planning* and those dealing with the *estimation of the cutting times and costing*.

OPERATION PLANNING

Specific Objective: *The student should be able to state the procedures involved in selecting a sequence of processes for the manufacture of given components.*

Machine selection

It is of little use proceeding with an operation layout if the work cannot be accommodated in the machine, or if the machine has the capacity for the work but it cannot hold the tolerances required. Equally, given a selection of machines all capable of doing the job, the smallest one with the required capacity should be chosen – it is not economic in engineering to use steam hammers to crack nuts.

To enable the most suitable machine for a given job to be chosen, machine tool builders supply *capacity charts* for their machines which show the slide movements and adjustments available. A comparison of the work with the capacity chart enables the suitability of the machine for the job to be readily decided.

At the same time it is necessary to decide whether or not the chosen machine is capable of working to the tolerances required. New or recently reconditioned machines will hold tighter tolerances than old and worn machines. A given machine turning off the cross-slide may hold tighter tolerances than operations carried out on the same machine working off the turret slide; roller boxes may hold closer tolerances than knee-tool holders working in the same turret. Thus, as well as knowing the *capacity* of the machine it is also necessary to know its *capability* for different types of operation. Quality control departments can provide this information by carrying out process capability studies. Such studies will be discussed in level 4 of this subject, and the methods used to obtain the information will not be considered here. Suffice it to say that, in well run quality-conscious organisations, the information is available and should be considered when selecting a machine for a given job.

Operation Layouts

The cost of a part often depends on whether the machine set-up is the most efficient for the part and, although in many cases the sequence of operations is self-evident, it is often necessary to prepare a tool layout. This is a plan-view of the part with all the tooling in position. It enables the planning engineer to decide precisely what tools are required and to ensure that they are available before work begins. A comparison with the capacity chart will show whether or not the machine will accommodate the job and when the machine and layout have been decided the operating time can be estimated.

The actual sequence of operations can best be arrived at by adopting a questioning attitude and for each part of the process asking the following questions which may be presented in the form of a check list:

1. Is machining necessary or can the maximum diameter of the part be left as bright bar stock?

2. Can a diameter be finish-machined in one pass, or are two cuts required, either for total metal removal or as roughing and finishing cuts?

3. Will the operation be carried out from the turret or cross-slide?

4. If the operation is being carried out from the turret, will it be best carried out with a roller box or a knee turning tool?

5. If a roller box is to be used will it be rollers leading or rollers following.

6. If a knee turning tool holder is to be used, can operations be combined?

7. Will cross-slide operations be plain-turning or will form-tools be required?

8. If two diameters are to be turned is it possible to turn the *smallest* diameter first to reduce the length to be turned for the larger diameter?

Fig. 5.11(a) is a tooling layout for a simple capstan operation to produce a bolt. If this

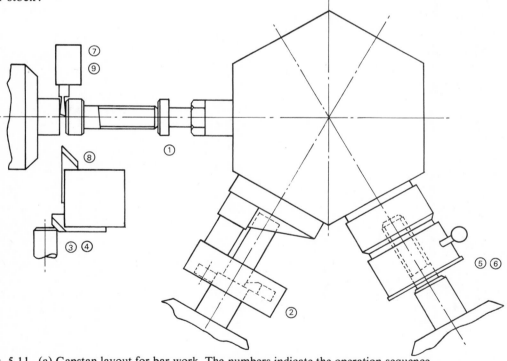

Fig. 5.11. (a) Capstan layout for bar work. The numbers indicate the operation sequence.

Part no. _____

Part name: Bolt

1 Operatn. No.	2 Operation	3 Tooling	4 Turret or X Slide	5 Speed m/Min	5 Speed Rev/Min	6 Feed rate cuts/cm	7 Total revs	8 Cutting time, min.	9 Non-productive time, min.	10 Remarks
1	Stock to stop	Bar stop	T	–	–	–	–	–	0.105	
	Index & lock turret	–	T	–	–	–	–	–	0.100	
2	Turn 25mm × 75mm long	Roller box	T	50	600	30	225	0.375		Rollers leading h.s.s. tool
	Index & lock turret	–	–	–	–	–	–	–	0.100	
	Position X slide & set stop	–	–	–	–	–	–	–	0.130	
3	Chamfer turned end	45° r.h. tool	X	50	600	hand	60	0.100		Front tool post
	Position X slide & set stop	–	–	–	–	–	–	–	0.130	Ready for op. 7
4	Chamfer head	45° r.h. tool	–	–	–	–	–	0.100		As for op. 3
	Change speed	–	–	–	–	–	–	–	0.100	
5	Rough thread	S/O die head	T	10	60	10	50	0.833		25mm × 1mm pitch chasers
	Withdraw & reset die head	–	–	–	–	–	–	–	0.080	
6	Finish thread	S/O die head	T	10	60	10	50	0.833		As for op. 5
	Index turret 3 stations	–	–	–	–	–	–	–	0.200	
	Position X slide & set stop	–	–	–	–	–	–	–	0.130	
	Change speed	–	–	–	–	–	–	–	0.080	
7	Partial part-off	Parting-off tool	X	25	150	hand	30	0.200		Ready for op. 7
	Index front tool post	–	–	–	–	–	–	–	0.110	
8	Chamfer head	45° l.h. tool	X	25	150	hand	15	0.100		Set relative to p/o tool – no slide adjustment
	Index front tool post	–	–	–	–	–	–	–	0.110	Ready for op. 3
9	Finish part-off	Parting-off tool	X	25	150	hand	30	0.200		As for op. 8
	Remove part	–	–	–	–	–	–	–	0.04	
	Change speed	–	–	–	–	–	–	–	0.100	Ready for op. 2
							Totals	2.741	1.515	Total time/part 4.256 min

Fig. 5.11 (b) Operation sheet for bar work.

layout is considered in the light of the above questions, it can be seen how they have been answered in this case.

1. No machining was necessary for the bolt head – it was left as bright bar.
2. The bolt diameter can be machined in one pass.
3. The turning operation is carried out from the turret using a roller box with rollers leading to give good concentricity with the head.
4. The chamfering operations are carried out using separate tools and operations. Operations 4 (chamfer front of head), 7 (partial part-off) and 8 (chamfer back of head) could have been combined in a single form-tool but in this case the quantities required did not warrant the cost of such a tool.
5. Threading the bolt required a roughing and finishing operation.

The operation sequence therefore became:

1. Feed stock to stop . . . *stop required*
2. Roller box diameter . . . *roller leading box*
3. Chamfer turned end . . . *X-slide chamfer tool*
4. Chamfer front of head . . . *same tool*
5. Rough thread . . . *die-head*
6. Finish thread . . . *same die-head*
7. Partial part-off . . . *rear X-slide parting tool*
8. Chamfer back of head . . . *X-slide chamfer tool*
9. Finish part-off . . . *rear X-slide – see (7)*

Similarly fig. 5.12 shows a tooling layout for a chucking operation in which knee-turning tools are used and operations have been combined. In both cases note that the tooling shown is stylised, rather than a true representation. Thus in fig. 5.11 the roller and tool in the roller box are in their correct positions and the overall dimensions of the box are correct but the full details are not shown. In fig. 5.12, the knee-turning tools are shown in the horizontal position rather than the vertical position normally used. This enables the cutting tools to be shown in the correct position relative to the work, but the knee-tool guide bushes are not shown.

The amount of detail shown is really a matter of individual taste, but the actual position of the cutting tools and overall dimensions of equipment should be correct.

ESTIMATION OF OPERATION TIME

Specific Objective: *The student should be able to calculate machining times and labour costs given appropriate machining conditions.*

The time required to machine a part depends upon the cutting speeds and feeds used for the various operations, and also upon the non-productive operations which, although vital, do not directly contribute to the value of the part, e.g., load and unload, change speed, feed stock to stop, index turret, etc. The actual cutting time can be accurately calculated from a knowledge of the cutting speeds and feeds used.

To illustrate the calculations of cutting times consider a length of bar of 25 mm diameter to be turned down to 20 mm diameter for a length of 75 mm, using a knee turning tool with tungsten-carbide tooling. The material is free-cutting mild steel, allowing a cutting speed of 150 m/min and a feed rate of 30 cuts/cm.

Note that the cutting speed is the peripheral or surface speed in metres per minute and the feed rate indicates the number of revolutions the work must make as the tool advances 1 cm.

Circumference of bar
$$= \pi \times \text{diameter}$$
$$= \pi \times 25 \text{ mm}$$
$$= 78.5 \text{ mm}$$

Work speed in revolutions per minute

$$= \frac{\text{Cutting speed in metres per minute}}{\text{Circumference in metres}}$$

$$= \frac{150 \text{ m/min}}{78.5/1000 \text{ m}}$$

$$= \frac{150 \times 1000 \text{ rev/min}}{78.5}$$

$$= 1910 \text{ rev/min}$$

143

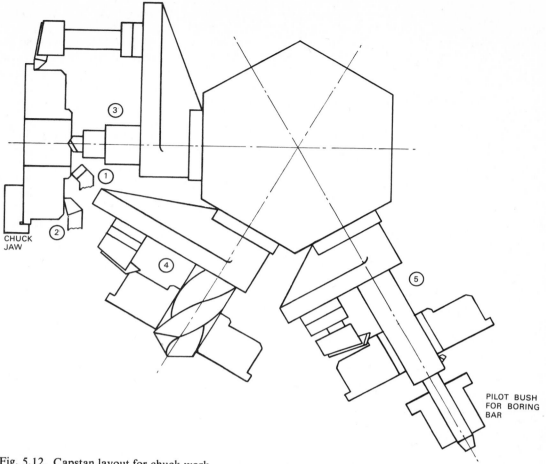

CHUCK
JAW

PILOT BUSH
FOR BORING
BAR

Fig. 5.12. Capstan layout for chuck work.

It is unlikely that the machine would have this precise speed available so for considerations of tool life the next nearest speed *downwards* is chosen. Assume a speed of 1800 rev/min.

Length of work
$$= 75 \text{ mm} = 7.5 \text{ cm}$$
Feed rate $= 30 \text{ cuts/cm}$
Number of revolutions
$$= \text{length in centimetres} \times \text{feed rate in cuts per centimetre}$$
$$= 7.5 \text{ cm} \times 30 \text{ cuts/cm}$$
$$= 225 \text{ rev}$$

Cutting time
$$= \frac{\text{No. of revolutions}}{\text{Speed in revolutions per minute}}$$
$$= \frac{225 \text{ rev}}{1800 \text{ rev/min}}$$
$$= 0.125 \text{ min}$$

If the calculations for a series of operations were carried out in the above manner, they would become hopelessly spread out and untidy. It is far better to set the calculations out in tabular form. The table can include the times for non-productive operations and a total floor-to-floor time can be arrived at. Consider the bolt shown in fig. 5.11(a), for which the table, or operation sheet, would be as shown in fig. 5.11(b).

Columns 1, 2, 3 and 4 are self-explanatory.

Column 5 shows the cutting speed to be used and from this the work speed in revolutions per minute is calculated.

Column 6 is the feed rate in cuts per centimetre, which enables the number of revolutions required for the operations to be calculated.

Column 7 is the number of revolutions required for the operation and is obtained by multiplying the feed rate in column 6 by the length of work to be cut.

Column 8 is the time required for the cutting operations. The total revolutions are known and are divided by the revolutions per minute to give the cutting time in minutes.

Column 9 shows the time required for non-productive operations. These non-productive times are known for different machines by work study engineers from past experience, those used here being for a medium-sized capstan lathe. A full list of them is not given but from the values used in fig. 5.11(b) the student could compile a useful basic table of such values. These he could augment by estimating times for other non-productive operations with which he is familiar.

Column 10 is a useful place to include any explanatory comments, as shown.

From the operation sheet, fig. 5.11(b), it would appear that a capstan operator, working under the conditions shown, could produce this part in 4.256 min, or 14.1 parts per hour or 112.8 such parts in an eight-hour shift. Allowances must be made, however, for unavoidable stoppages. Since the part is 100 mm long plus a parting-off allowance, about 105 mm is required per part. Thus, if the operator is using bars of 3 m length he must stop after every 28 parts and insert a new length of bar. Time must be allowed for this. Allowance must also be made for tool maintenance and for the operator to attend to the needs of

nature. Unfortunately, these two contingencies rarely coincide.

Various organisations have their own arrangements and standards for these allowances but detailed descriptions of them form an important part of work study and would be out of place in this book. However, the student should be aware of the need for such allowances, and that the simple time-estimation sheet is not the whole story.

Assume that in a particular organisation, personal allowances total 5 min/hour, and that a 10% allowance for tool maintenance is added to the machining time. Further, a 10% allowance is given for unforeseen contingencies and it takes the operator 1.00 minutes to change the bar stock. From the operation sheet:

$$
\begin{aligned}
\text{Machining time per part} &= 4.26 \text{ min *} \\
10\% \text{ tool maintenance} &= \underline{0.43 \text{ min}} \\
\text{Running total} &= \underline{4.69 \text{ min}} \\
10\% \text{ contingencies allowance} &= \underline{0.47 \text{ min}} \\
\textit{Total machining time/part} &= \underline{5.16 \text{ min}}
\end{aligned}
$$

(Rounded to decimal minutes)

The bar stock requires replacement after 28 parts have been produced. Based on an average machining time of 5.16 min/part then:

Time between bar stock replacements
$$
\begin{aligned}
&= 5.16 \times 28 \\
&= 144.48 \text{ min}
\end{aligned}
$$

In a shift length of 480 min then,

Number of bar replacements $= \dfrac{480}{144.48} = 3.32$

As the operator will not replace 0.32 of a bar, this is rounded up to 4 bar replacements in an 8 hour shift.

$$
\begin{aligned}
\text{Shift length} &= 480 \text{ min} \\
\begin{matrix}5 \text{ min/hour per-}\\ \text{sonal allowance}\end{matrix} &= \underline{\ \ 40 \text{ min}} \\
&\ \ \ \underline{440 \text{ min}}
\end{aligned}
$$

$$
\begin{aligned}
\begin{matrix}4 \text{ min bar change}\\ \text{allowance}\end{matrix} &= \underline{\ \ \ 4 \text{ min}} \\
\begin{matrix}\text{Actual machining}\\ \text{time}\end{matrix} &= \underline{436 \text{ min}}
\end{aligned}
$$

$$\text{Production/shift} = \frac{\text{Machining time available}}{\text{Machining time/part}}$$

$$= \frac{436 \text{ min}}{5.16 \text{ min/part}}$$

$$= \underline{84.5 \text{ parts/shift}}$$

This must now be compared with the earnings of the operator to obtain the machining cost per part. Note that, although an operator cannot produce the odd 0.5 parts per shift, in the long run it can be reasonably expected that production will average out at 84.5 parts/shift and thus the calculated figure is used.

Assuming that the operator is paid £2.00 per hour, then in an 8 hour shift he will have earned £16.00 and therefore:

$$\text{Machining costs/part} = \frac{£16.00}{84.5}$$

$$= \underline{18.93 \text{ pence/part}}$$

It must be noted that this is not the total cost of the part — it is only the machining cost. Obviously the material cost must be added and this gives the total materials and labour cost. However there is a great deal more to be added yet. The shop chargehand and the area foreman have added nothing to the value of the part but they have to be paid from the income gained by the part when it is sold. So must the wages department, the sales department, the planning department, the drawing office and so on. Factory rent, heating, lighting and other services must be paid for. The machine on which the part is made has been paid for, it depreciates and this must be allowed for as well as provision being made for replacing the machine when it is finally scrapped. All of these things are often lumped together under the heading of factory overheads, and a reasonable estimate for the selling price at the factory is: (materials + labour) + 400%, or five times the materials and labour costs. This would make the total cost of the bolt about £1.00. This, of course, seems outrageous, but it must be remembered that standard bolts are not made on capstan lathes — special machines are used which produce them in about 5 sec, not 5 min. The only bolts of this type which would be produced on a capstan lathe would be small batches of special ones in non-standard sizes. For these the price is, perhaps, more reasonable.

QUESTIONS

1. During a metal-cutting operation with a single point tool, the measured forces acting on the tool are found to be:
   Radial Force = 200 kN
   Axial Force = 700 kN
   Tangential Force = 1100 kN
Calculate the resultant force acting on the tool and its direction relative to the horizontal and to the machine axis.

2. Consider the forces in the horizontal plane in question 1, and suggest a value for the plan approach angle for the tool.

   What would be the effect of increasing the plan approach angle?

3. In a cutting test for a roughing cut on mild steel using a tungsten carbide tool, a cutting speed of 126 metres/min gave a tool life of 20 min, and a cutting speed of 134 metres/min gave a tool life of 15 min.

   What cutting speed should be used to give a tool life of 60 min.

4. A cutting test was carried out to measure the variation in tangential force on a cutting tool with a change in cutting speed. The values obtained are tabulated below for a depth of cut of 8 mm on a bar 120 mm diameter, with a feed rate of 0.2 mm/rev.

   Cutting speed (metre/min) 5.88 7.66 10.6 14.14 18.9 24.75 33.58 44.76 61.3 83.7

   Tangential force (N) 3700 3600 3100 3000 2700 2400 2275 2240 2275 2400

   Calculate, for each speed, the power absorbed at the tool and the specific rate of metal removal (mm³/min. kW). Plot a graph of specific rate of metal removal against cutting speed and hence determine the optimum speed for this operation.

5. Make a sketch of a negative-rake clamped-tip tool showing clearly:
   (a) how the tip is supported;
   (b) how the tip is located;
   (c) how the tip is clamped.
   What are the advantages of this form of holder over brazed tips for tungsten carbide tools?

6. Suggest suitable tooling for the following turning operations:
   (a) Taking shallow cuts on free-cutting mild steel using a roller-box on an old capstan lathe, having maximum spindle speed of 500 rev/min. The bar diameter is 20 mm.
   (b) Machining a 100 mm diameter bore in an aluminium casting on a modern machine, with a maximum spindle speed of 5000 rev/min.
   In each case give reasons for your answer.

7. Examine a capstan lathe to which you have access and:
   (a) make a capacity chart;
   (b) tabulate the speeds and feeds available;
   (c) estimate times for non-cutting operations.

8. Make a capstan layout and operations sheet for producing the part shown in fig. 5.13 from bar stock of 50 mm diameter.
   (a) Estimate the basic machining time.
   (b) Assuming suitable allowances estimate the parts produced in an eight hour shift.

ANSWERS

1. $R = 1319$ kN at 15.95° to axis and 56.5° to horizontal.
3. $n = 0.214$; $C = 239$; 99.5 metre/min.

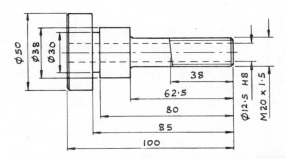

ALL DIMENSIONS IN mm

Fig. 5.13. Question 8.

147

6 Metal Cutting with Multi-point Tools

General Objective: *The student should recognise the principles of multi-point metal cutting as applied to milling cutters.*

MILLING CUTTERS

The tooth of a milling cutter can be likened to a boring tool in that both work on the inside of a curved surface. A major difference for milling cutters of disc type when they are cutting a slot is that the chips cannot escape until the tooth in front of them is well clear of the slot. Thus the space between the teeth is important since it must adequately contain the chip produced. Secondary clearance is provided to increase this space and other angles are similar to those of a boring tool.

For a given size of cutter the number and size of teeth depend upon:

1. tooth load – what will be the maximum load the tooth must bear in service?

2. chip clearance – will the space between the teeth contain the chip produced at maximum load?

The two factors depend upon each other. For instance, if the chip clearance is too small it can be increased either:

(a) by retaining the same tooth size but reducing the number of teeth. For a given depth of cut and rate of feed, each tooth then has to do more work and the load on each tooth is increased; or

(b) by retaining the same number and depth of teeth but increasing the gap by making the tooth thinner. This reduces the load which the tooth can take and means a lower maximum depth of cut and feed, but these in turn will mean that less chip space is required.

Thus the size, shape and number of teeth are a carefully thought-out compromise between tooth strength and chip space, and the necessary proportions must be maintained. If a cutter is ground excessively on the clearance face so that the tooth space is considerably reduced, the cutter should be re-gashed to restore the original tooth shape and give adequate chip space.

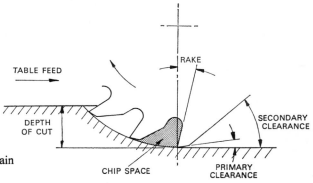

Fig. 6.1. Teeth of a milling cutter showing main features.

TYPES OF MILLING CUTTER

1. *Slitting saws and slot cutters*

These cutters have no side-cutting teeth and cut only on the periphery. To give side clearance and prevent binding in the slot which they cut, they are dished, or made wider at the cutting edge than at the centre, as shown in fig. 6.2. Note that the dishing serves the same purpose as the front-to-back clearance on a parting-off tool for a lathe.

It is important that there should be no side thrust on these cutters; they should not be used to cut on one side but only in a slot, to give a balanced load.

2. *Side-and-face milling cutters*

These cutters have side cutting edges as well as the normal front cutting edges. Consequently they can be used to cut on the side as well as on the face. They are usually wider than slot cutters and more able to withstand side loads. Fig. 6.3 shows such a cutter and that the side cutting edges have

1. zero rake angle;
2. little chip space.

These faults can be overcome by removing alternate side-cutting teeth on opposite sides. This increases the chip clearance and allows the teeth to be angled in opposite directions to give rake to the side-cutting edges, producing the staggered-tooth side-and-face cutter shown in fig. 6.4.

3. *Slab or roller milling cutters*

These are large cylindrical cutters used for machining horizontal surfaces. It should be noted that the shape of the workface is determined by the shape of the cutter, and if the teeth do not lie on a true cylinder the work will not be milled flat or parallel.

The teeth may be straight or helical but helical teeth are preferred. When straight teeth are used, each tooth bites into the work with its complete width, creating shock loads and vibrations, and the chips pile up on the work face in front of the cutter.

A cylindrical cutter with a steep helix angle is shown in fig. 6.5. The end elevation and front elevation show the general appearance of the cutter, but the plan view is a view of the work during cutting with the lines of the teeth superimposed, and shows two effects of such a cutter.

The forces involved in plan are shown. They are

Fn, the force normal to the tooth;
Ft, the force tangential to the cutter;
Fe, the end thrust.

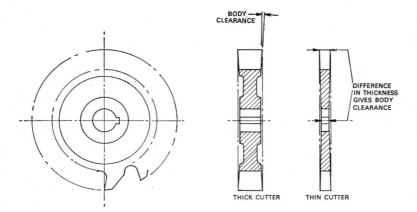

Fig. 6.2. Slot milling cutters. Note how body clearance is obtained and compare with parting-off tools.

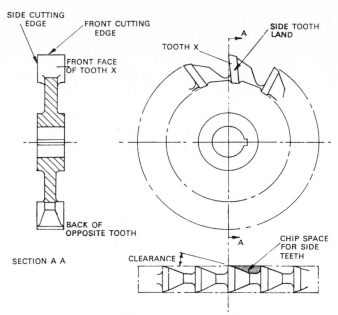

Fig. 6.3. Side-and-face cutter. Note that the slide cutting edges have zero rake. Plan view. Note this is not a true plan, but a view of a straight line of teeth as if the cutter had been unrolled.

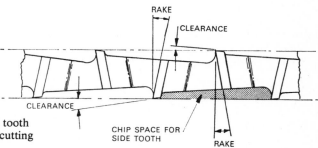

Fig. 6.4. Extended plan view of staggered tooth side-and-face cutter showing rake on side cutting teeth.

The lines of the teeth show that the cut is taken up and relinquished gradually by each tooth. Compared with a straight-toothed cutter, the number of teeth in operation is increased and the length of tooth in operation is reduced. At A, a tooth is starting to cut at a point; at BC it has progressed to its maximum width of cut and at DE the cut has started to reduce, to run out gradually at F. All these factors produce a much smoother cutting action, a better surface finish and a more even distribution of tooth load.

It should be noted that the action produces an end thrust on the cutter, and the steeper the helix the greater will be this end thrust. The cutter should therefore be mounted on the machine so that the end thrust is *towards* the body of the machine, and not tending to pull the arbor out of the machine spindle.

150

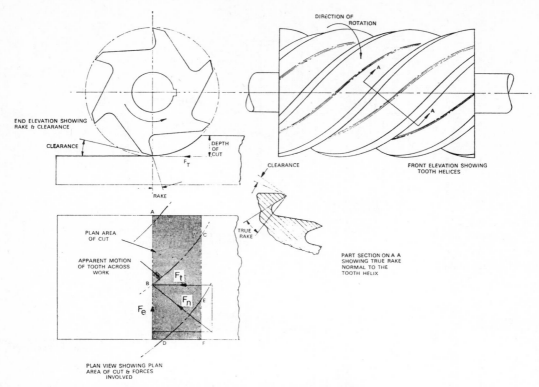

Fig. 6.5. Slab or rolling milling cutter and forces set up during its action.

4. *Face milling*

The width of end face that can be machined by a side-and-face cutter is limited by the diameter of the cutter and the fact that the arbor may foul the work, as shown in fig. 6.6(a). This can be overcome by putting the cutter on the end of the arbor, so to speak, and recessing its clamping arrangement. At the same time the cutter can be made broader and more able to support thrust loads.

As the cutter will be used only on one end, the teeth can be given a rake angle by making the flutes helical as shown in fig. 6.6(b). The angle of the helix gives rake angle to the teeth on the end of the cutter, not to those on the periphery.

5. *End mills and shell end mills*

These cutters may be relatively small and made from solid bar, when they are known as *end mills*. With larger cutters the cost of a solid cutter and shank, all of high-speed steel, would be prohibitive and such cutters are therefore made separately from their shanks, which may be made of plain carbon steel. Since they are shaped like a cylindrical shell, they are called *shell end mills* and are mounted on their shanks as shown in fig. 6.7. Note that the shank takes the place of the usual arbor in the machine spindle and is positively driven by tenons, and that the cutter is located on the shank by a spigot and in turn has a positive drive by other tenons. Fig. 6.8 shows three small solid-shank end mills, all with the same direction of rotation but with different helix angles. Note the effect of the helix on the rake angle of the end cutting teeth.

151

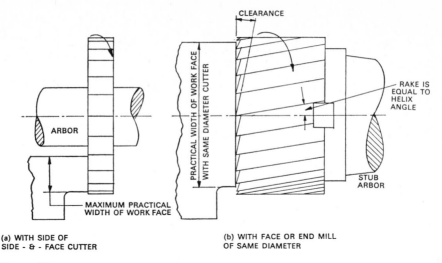

(a) WITH SIDE OF
SIDE - & - FACE CUTTER

(b) WITH FACE OR END MILL
OF SAME DIAMETER

Fig. 6.6. Face milling.

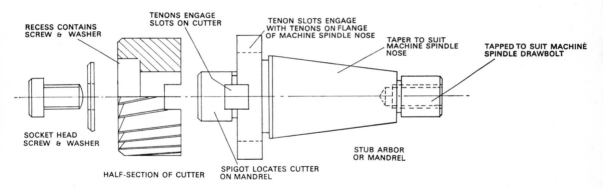

Fig. 6.7. Exploded assembly of shell end milling cutter and arbor.

The *slot drill* is a particular form of end mill having two flutes only and cutting teeth on the end, not on the periphery, as on the end mill. Thus it can only cut with its end teeth and not with the peripheral teeth. Whereas an end mill can be used to cut a step in the side of a block by taking a series of shallow cuts with its peripheral teeth a slot drill cannot. The sole purpose of the slot drill is for milling keyways and shallow slots and, having two teeth only, each tooth has a large amount of chip clearance

and the cutter does not tend to clog in the slot with the chips it produces.

A feature of end mills and slot drills is that they can be traversed through the work in any direction at right angles to the cutter axis. For instance, with an end mill mounted vertically and sunk into the work by raising the table, an 'L'-shaped slot can be produced simply by first traversing the table longitudinally and then transversely. Similarly if the work is mounted on a rotary table and the cutter sunk into the

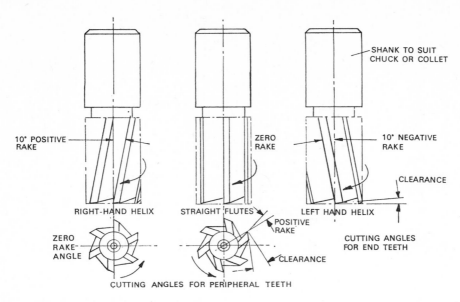

Fig. 6.8. Types of end mill and their cutting angles.

work, a slot can be produced in the form of an arc of a circle, simply by rotating the table. This cannot be done with peripheral type cutters such as the side and face cutter. For this reason the end mill is much more versatile than peripheral-type cutters. It is interesting to note that, in almost all cases, numerically-controlled milling machines, which will be considered at a later stage and which are ideal for complex profiling, use end mills.

6. *Inserted tooth cutters*
In Chapter 5 the development of clamped-tip type cutting tools was discussed. A similar development has taken place with milling cutters. Until recently, inserted-tooth face mills had a series of brazed tips set into the body as shown in fig. 6.9(a). When the tips were worn, the regrinding and resetting of the teeth were critical to get all the tip points at the same height. Modern cutters employ throw-away tips which are accurately located in the body and, when replaced, position themselves accurately. Such a cutter is shown in fig. 6.9(b). These milling cutters, usually face mills, are used at

much higher speeds than conventional cutters, with a fine feed, to give very high rates of metal removal. As with turning tools, machine rigidity, power and speed are extremely important and these milling cutters should only be used under suitable conditions, not on an old, underpowered machine just for the sake of using this type of tool. Initially these developments were restricted to face mills but, as can be seen from fig. 6.10, other types of cutters are now obtainable with clamped indexable tips for work other than face milling.

MORE SPECIALISED CUTTERS AND OPERATIONS

1. *Form-relieved cutters*
All the examples described have dealt with flat surfaces, except for the use of the rotary table to generate an arc. However, some shapes or forms need to be copied by specially-made cutters, such as that shown in fig. 6.11. These cutters are usually form-relieved, i.e. the form of each tooth is continued and preserved for the full depth of the tooth, clearance being obtained

153

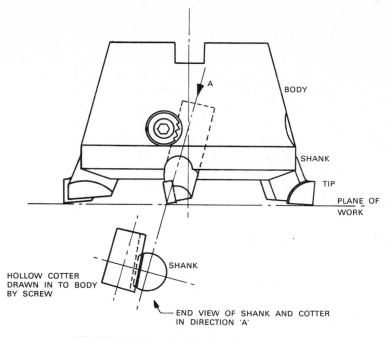

Fig. 6.9. (a) Face milling cutter with negative rake tips brazed to shanks.

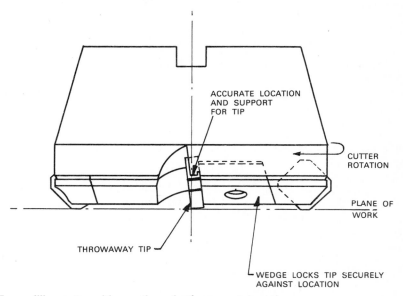

Fig. 6.9. (b) Face milling cutter with negative-rake throwaway-type tips.

154

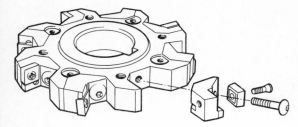

Fig. 6.10. Staggered tooth milling cutter with indexable inserted teeth. Courtesy of Madison Tools Ltd.

by making each tooth a small part of a spiral whose cross-section is the form of the tooth. Thus however far the teeth are ground back on the rake face their form will remain correct. This is true only if the rake angle is correct and care must be exercised in sharpening such cutters to preserve the correct rake angle.

Form relieving is used for all types of profile cutters, including radius cutters and those used for producing special profiles.

2. *Dovetail cutters*
A dovetail cannot be produced by a normal right-angled cutter and a special type of cutter is

used, as shown in fig. 6.12. It machines the base surface with its end teeth and the angled surfaces with its inclined cutting face. Similar single and double angle cutters may be mounted on the machine arbor for use in the horizontal plane.

3. *Tee-slot cutters*
Tee-slots are first machined by cutting the vertical slot to full depth with an end mill or slot drill. The tee of the slot is then cut by a special cutter as shown in fig. 6.13. It consists essentially of a small side-and-face cutter mounted on a shank, with a neck small enough to clear the vertical slot machined in the first operation. With the aid of a rotary table, a tee-slot can be machined to an arc of a circle.

4. *Woodruff cutters*
These cutters are used for producing Woodruff key-seats and must conform to the shape of the key, which is a segment of a circle. The cutter is too small to be mounted on the machine arbor and it is made in the form of a slot cutter on a solid shank, side clearance being obtained by dishing. Unlike a tee-slot cutter it has no side-cutting teeth and can further sometimes be

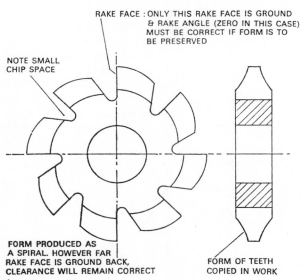

RAKE FACE : ONLY THIS RAKE FACE IS GROUND & RAKE ANGLE (ZERO IN THIS CASE) MUST BE CORRECT IF FORM IS TO BE PRESERVED

NOTE SMALL CHIP SPACE

FORM PRODUCED AS A SPIRAL. HOWEVER FAR RAKE FACE IS GROUND BACK, CLEARANCE WILL REMAIN CORRECT

FORM OF TEETH COPIED IN WORK

Fig. 6.11. Form-relieved cutter.

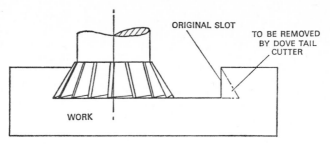

Fig. 6.12. Action of dovetail cutter.

identified by the fact that its end is not flat but carries a centred spigot which allows a support centre to be used as shown in fig. 6.14. To produce the key-seat, the cutter is positioned over the work and sunk directly to the required depth. Different sizes of Woodruff keys, and therefore of Woodruff cutters, are used.

UP-CUT AND DOWN-CUT MILLING

Specific Objective: *The student should be able to define up-cut and down-cut milling and state the merits of each process.*

In the work devoted to metal cutting discussed in Levels 1 and 2 of this series, the accent has been on the fundamental requirements of metal cutting. At this stage the student should be

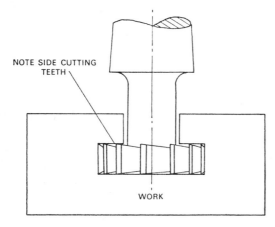

Fig. 6.13. Action of tee-slot cutter.

becoming more aware of the importance of high metal removal rates for economic production and this is equally true of milling as well as the continuous cutting turning operations. Using peripheral cutters under certain conditions, the metal removal rate can be improved by reversing the cutter/work relationship. Thus instead of using conventional, or up-cut milling, the cutter is reversed on its arbor, its direction of rotation is reversed but the direction of feed remains the same. This is called *down-cut milling*, or in some cases, *climb-cutting*.[1]

The relative motions of the cutter and work are shown in fig 6.15(a) and (b) for both up-cut and down-cut milling. An examination of the forces involved shows an immediate difference in the process. If F_t is the force normal to the tooth, i.e., tangential to the cutter, the horizontal component F_h is in opposition to the direction of feed and the vertical component F_v is upwards in the case of up-cutting. Thus the horizontal component, operating against the leadscrew, takes up any backlash while the vertical component tends to lift the work off the table against the clamping force.

In down-cut milling, the horizontal force is in the direction of the feed motion and any backlash in the leadscrew will allow the work to be dragged into the cutter. This effectively increases the feed per tooth causing overloading of the cutter and probably breakage. This

[1] The term *climb-cutting* is more usually applied to hobbing, a gear manufacturing process in which the work-spindle is vertical and the cutter appears to 'climb' up the work.

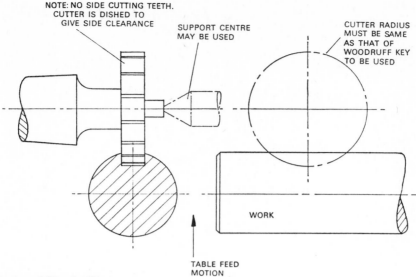

NOTE: NO SIDE CUTTING TEETH.
CUTTER IS DISHED TO
GIVE SIDE CLEARANCE

SUPPORT CENTRE
MAY BE USED

CUTTER RADIUS
MUST BE SAME
AS THAT OF
WOODRUFF KEY
TO BE USED

WORK

TABLE FEED
MOTION

Fig. 6.14. Action of Woodruff key-seat cutter.

would appear to eliminate down-cut milling as a method of cutting but, because of other advantages, machines have been developed in which the leadscrew backlash can be eliminated, the cost of the mechanism required being offset by the increased metal removal rates possible. Note also that the vertical force F_v is downwards, thus assisting the work-clamping forces.

One reason for the increased metal removal rates is shown in fig. 6.15. The rate at which metal can be removed depends on the cutting speed and feed rate used, and an increase in these will contribute to reduced tool life. In up-cut milling, at the start of the cut the depth of cut is zero, rubbing tends to occur and causes a high rate of cutter wear. In down-cut milling, however, by the time a tooth engages the work the table has fed forwards, giving a definite cut as the tooth engages the work and a gradual runout. This reduces the rate of cutter wear, enabling higher speeds and feeds to be used between cutter regrinds. The author has no figures available for milling but for hobbing, a process with a similar cutting action, increases of 20% in cutting speed plus 10% in feed rate, with an increased tool life of the order of 30%,

have been reported, with an improvement in surface finish.

Specific Objective: *The student should be able to derive a comparative shape of chip produced in up-cut and down-cut milling.*

If the shape of the chips produced by up-cut and down-cut milling are compared for similar cutting conditions, i.e. depth of cut and feed rate, a further reason for the improved cutting action becomes apparent. Fig. 6.16(a) represents the paths traversed by successive teeth of a milling cutter being used conventionally and the shaded area is the chip-shape produced. To generate this shape the procedure is as follows:

1. Draw a line XY parallel to the work surface representing the path of the cutter centre relative to the work.
2. Set off equi-spaced positions 1, 2, 3, . . . 10, representing positions of the cutter centre during the path of the tooth through the work.
3. From point 1, draw a vertical line to the position A where the tooth will enter the work. Distance A1 represents the cutter radius.

157

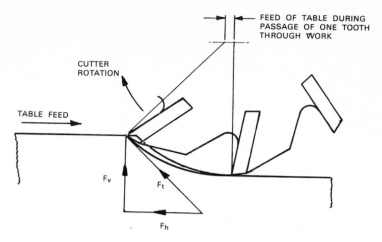

Fig. 6.15. (a) Up-cut milling.

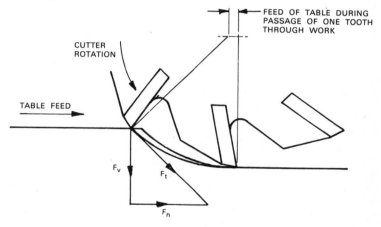

Fig. 6.15. (b) Down-cut milling.

4. From point 10, strike an arc equal to the cutter radius to intersect the unfinished work surface at K. The angle between 10K and the vertical is the angle turned through by the tooth in the time the machine table has moved from position 1 to position 10. Let this angle be $X°$.

5. Divide this angle by 10 and the angle thus obtained is the angle through which the cutter rotates in moving one position along line AB.

6. Set off line 2B at $\dfrac{X°}{10}$ and strike off the cutter radius to give point B.

7. Set off line 3C at $\dfrac{2X°}{10}$ and strike off the cutter radius to give point C and so on.

8. Draw a smooth curve through points A, B, C, ... K. This will be the path cut through the metal by one tooth. If the next tooth starts to cut when the cutter centre has reached point 3, then a similar curve may be generated for the next tooth starting vertically under point 3.

The zone shaded between these curves is the chip shape for up-cut milling for the cutter radius, depth of cut and feed/tooth used. Fig.

158

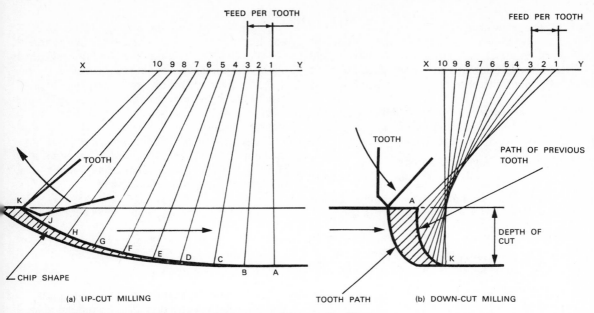

Fig. 6.16. Comparative chip shapes for up-cut and down-cut milling.

6.16(b) can be developed using a similar procedure for down-cut milling, but in this case line A1 is at the angle used for K10 in the previous case, line K10 in this case is vertical and the intermediate lines are drawn *back* at $\frac{X°}{10}$, $\frac{2X°}{10}$, $\frac{3X°}{10}$, etc., as shown. Again the feed/tooth is the distance between points 1 and 3 and if a similar curve is generated starting from point 3 then the shaded area between the curves is the chip shape for down-cut milling.

The difference between the chip shapes is immediately apparent. That for down-cut milling gives a much shorter chip of more nearly equal thickness throughout the tooth passage.

It must be emphasised that only machines on which backlash can be eliminated from the table feed mechanism should be used for down-cut milling. Machines with a hydraulic table feed are eminently suitable but those with the leadscrew-and-nut type of feed drive must be fitted with a backlash eliminator, or serious damage may be done to the cutter.

OPTIMUM CUTTING CONDITIONS

Specific Objective: *The student should be able to select rates of speed and feed for given cutters.*

For any cutting process on a given machine there is a metal removal rate which is the most economic that can be achieved under the prevailing conditions. It will depend on:

(a) Tool life between regrinds;
(b) Power available.

(c) *Tool life*
We have seen in Chapter 3 that the tool life T is related to the cutting speed V by a law of the form

$VT^n = C$ where n and C are constants depending upon the conditions.

Thus, if we require a certain tool life we can ascertain the cutting speed required to give that life.

159

(b) *Power available*

It can be shown that under given conditions of work material and cutting tool material the volume of work material which can be removed per minute with one kilowatt of power is almost constant.

Let the width of cut be W millimetres, the depth of cut be d millimetres and the feed rate be f millimetres per minute. Then:

Volume removed per minute
$$= W \times d \times f \text{ cubic millimetres}$$

If the volume removed per minute for the expenditure of 1 kW is the constant K:

Power required, P,

$$= \frac{\text{volume to be removed per minute}}{\text{volume removed per minute per kilowatt}}$$

$$P = \frac{Wdf}{K} \text{ kilowatts}$$

As the width of cut, W millimetres, is usually fixed either by the width of the work or the width of the cutter, and the depth of cut is usually such that we wish to remove the metal at one pass, it follows that f, the feed rate in millimetres per minute, must be calculated to allow this to be done with the power available, and the power required, P kilowatts, must not exceed this.

Therefore feed rate $f = \dfrac{P \times K}{W \times d}$ mm per minute

Approximate values of K in cubic millimetres per minute per kilowatt of power supply are given in the following table. It must be emphasised that they are average values of K and in practice they will vary according to the condition of the machine, the condition of the cutter, the coolant used and other factors.

For a given operation the actual value of K can be found by performing a test on a milling machine with a wattmeter wired into the machine circuit.

Consider now the face milling of a piece of medium-carbon steel 100 mm wide with a H.S.S. cutter, the depth of cut being 5 mm and the optimum power available 5 kW.

Feed rate $f = \dfrac{KP}{Wd}$ millimetres per minute

$$= \frac{20\,000 \times 5}{100 \times 5}$$

$$\therefore f = 200 \text{ mm/min}$$

It has been found that to give the required tool life a cutting speed of 20 m/min is suitable for a cutter of 120 mm diameter.

Spindle speed in revolutions per minute

$$= \frac{\text{cutting speed in metres per minute}}{\text{circumference of cutter in metres}}$$

$$= \frac{20 \text{ m/min}}{\pi \times 120/1\,000 \text{ m}}$$

Spindle speed
$$= \quad 53 \text{ rev/min}$$

Feed in millimetres per revolution

$$= \quad \frac{200 \text{ mm/min}}{53 \text{ rev/min}}$$

$$= \quad 3.75 \text{ mm/rev}$$

Work material	Cutter material	Type of cutter	K mm^3/min.kW
Medium carbon steel	H.S.S.	Slab mill	20 000
	H.S.S.	Face mill	35 000
	Tungsten carbide	Negative-rake face mill	20 000
Cast iron	H.S.S.	Slab mill	40 000
	H.S.S.	Face mill	60 000

If the cutter has 15 teeth, then the feed per tooth can be found as:

feed per tooth

$$= \frac{3.75 \text{ mm/rev}}{15 \text{ teeth/rev}}$$

$$= 0.25 \text{ mm/tooth}$$

This would be suitable for roughing, which is required, since we are concerned with optimum rates of metal removal.

Another situation for a similar calculation would be to ensure that a large surface could be finish machined at one pass, using a large enough cutter. Alternatively, if a lot of material had to be removed and the required finish demanded a feed of no more than 0.06 mm/tooth, what would be the maximum allowable depth of cut? Could the surface be machined in one cut or would a roughing cut followed by a fine-feed finishing cut be preferable?

This type of calculation need not be restricted to milling operations. By experiment values of K could be found for turning operations and drilling processes, in fact for any metal-cutting process where high metal removal rates are required.

CALCULATION OF CUTTING TIMES

Examination of a milling machine shows that, unlike the lathe, the table feed rate is independent of the spindle speed. On a lathe, the feed rate is usually quoted in cuts per centimetre, i.e., the number of revolutions the spindle makes for a centimetre of traverse. On a milling machine, the table feed-rate is given directly in millimetres per minute irrespective of the spindle speed. Thus it would appear that a workpiece 200 mm long, with a table traverse speed of 50 mm/min, would take 4 min to machine. This, however, does not take into account the fact that a milling cutter does not attain its full depth of cut immediately it makes contact with the work. A distance x is required for the cut to build up at the start. This can be seen in fig 6.17 for peripheral milling, i.e., using a slab mill, slot cutter, or side-and-face cutter, and in fig. 6.18 for face milling.

Consider fig. 6.17.

In triangle OAB,
    OA = cutter radius R
    AB = approach factor x
    OB = $R - d$, where d = depth of cut

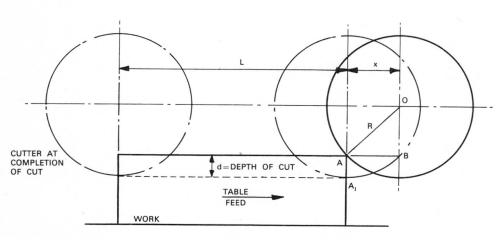

Fig. 6.17. Approach factor for peripheral milling. Cutter makes contact at A and reaches full depth of cut at A_1.

Using the theorem of Pythagoras,
$$(OA)^2 = (OB)^2 + (AB)^2$$
$$(AB)^2 = (OA)^2 - (OB)^2$$
$$\begin{aligned}x^2 &= R^2 - (R-d)^2\\ &= R^2 - (R^2 - 2Rd + d^2)\\ &= R^2 - R^2 + 2Rd - d^2\\ &= 2Rd - d^2\\ &= d(2R - d)\end{aligned}$$
but $2R =$ cutter diameter D
$$\therefore\ x^2 = d(D - d)$$

and approach factor $x = \sqrt{[d(D-d)]}$ where $d =$ depth of cut and $D =$ cutter diameter.

The full length of traverse is therefore $L + x$. It might appear that it should be $L + 2x$, an allowance being made for the cutter to clear the work at the end. This is not so, because once the spindle centre line has passed the edge of the work all the metal has been removed.

The conditions for face milling are shown in fig. 6.18, in which x is again the approach factor and R the radius of the cutter. In this case the width of the work, W, must be considered.

From fig. 6.18,

$$x = AB = OB - OA$$
But $OB = R$, the cutter radius.

In triangle OAC
$$OC^2 = OA^2 + AC^2$$
$$OA^2 = OC^2 - AC^2$$

But OC also $= R$

and $AC = \dfrac{W}{2}$

$$\therefore\ OA^2 = R^2 - \frac{W^2}{4}$$

$$OA = \sqrt{(R^2 - W^2/4)}$$

and $x = OB - OA$
$$= R - \sqrt{(R^2 - W^2/4)}$$

But $D =$ cutter diameter, then

$$x = \frac{D}{2} - \sqrt{\left(\frac{D^2}{4} - \frac{W^2}{4}\right)}$$
$$= \frac{D}{2} - \frac{1}{2}\sqrt{\left(D^2 - W^2\right)}$$

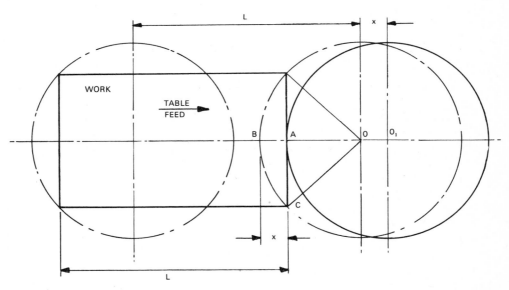

Fig. 6.18. Approach factor for face milling. Cutter makes contact at A with centre at O and reaches full width of cut when centre reaches O_1.

and therefore

Approach factor $x = \frac{1}{2}[D - \sqrt{(D^2 - W^2)}]$

Again no allowance need be made for run-out, indeed the cutter should be cleared away from the work as rapidly as possible to avoid rubbing. In fact, in face milling the spindle can be inclined so that the rear of the cutter clears the work by about 0·05 mm to avoid rubbing.

In both cases, then, the total traverse is given by

Total traverse $= (L + x)$ millimetres
where $L =$ work length (millimetres)
$x =$ approach factor (millimetres)

and

cutting time $= \dfrac{L + x}{f}$ minutes

$f =$ feed rate in millimetres per minute.

CUTTER GRINDING

The grinding of cutting tools and especially of milling cutters is a specialised process for which special grinding machines have been developed. A common type is one in which the tabie has longitudinal slides and cross slides as shown in fig 6.19. The double-sided wheel-head is mounted on a vertical column, which also carries the motor, so that the head can be raised, lowered and swivelled without any complicated driving arrangement. There is usually no power feed to the table motions but the hand controls are duplicated so that the operator can work from either the front or rear of the machine.

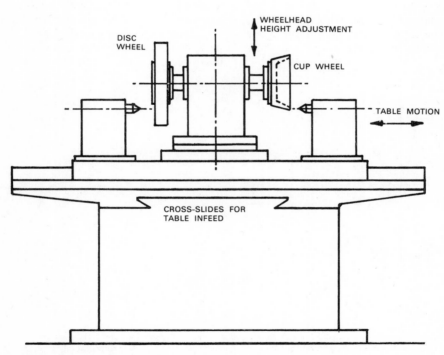

Fig. 6.19. Block layout of cutter grinding machine.

GRINDING MILLING CUTTERS

Specific Objective: *The student should be able to describe, with the aid of sketches, the principles of regrinding milling cutters.*

If a milling cutter is to be sharpened, a set-up must be arranged so that each tooth can be passed across the grinding wheel in turn, with the same amount of material being removed from each tooth. This necessitates a method of indexing. At the same time, the set-up must be so arranged that the correct angle is produced on the cutter, to give the correct cutting action when the cutter is in use.

The method of indexing used depends on the type of cutter being ground, and the relative positioning of the cutter and wheel produces the correct cutting angles.

Whatever type of cutter is being ground, it is advisable to take a series of only light cuts, the same cut being taken on all teeth before feeding in the cutter to remove further material. If the cut is too heavy, local overheating and possible damage to the cutter may occur and the cutter will then require regrinding again in a short time.

METHOD OF INDEXING THE CUTTER

The cutter must not rotate during the grinding of each tooth, and rotation is prevented by locating one of the teeth against a metal stop or *tooth rest*. The cutter is held against this tooth rest by hand pressure and the tooth rest is often spring-loaded to facilitate the indexing of the cutter for the grinding of successive teeth. The tooth rest is generally positioned so that the force due to the grinding wheel rotation holds the cutter-tooth against it. This is not always possible and figs 6.20(a) and (b) show the two conditions. In fig. 6.20(a) the clearance face of a side-and-face cutter is being ground; the tooth rest locates on the tooth being ground and the rotation of the grinding wheel helps to hold the cutter against it. In fig. 6.20(b) the rake face of a cutter is being ground and the cutter must be held against the tooth rest by hand pressure, against the torque exerted by the pressure of the grinding wheel.

It should be noted that in both cases the tooth rest locates against the rake face of the tooth. If the cutter is badly worn this may be neither possible nor desirable, in which case a separate index plate is mounted on the cutter mandrel

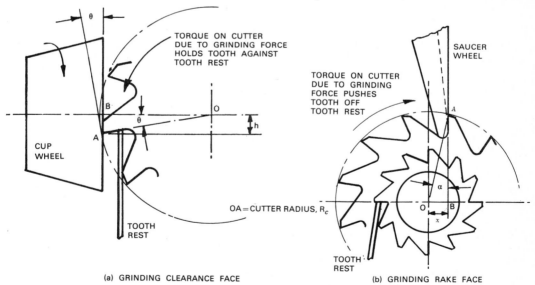

(a) GRINDING CLEARANCE FACE   (b) GRINDING RAKE FACE

Fig. 6.20. Cutter grinding.

and the tooth rest locates in the notches of the index plate.

If the end teeth of an end mill or shell end mill are to be reground it will be necessary for the cutter to overhang and be inclined, as shown in fig. 6.21, in which case it is mounted in a special fixture whose axis can be given the necessary inclination.

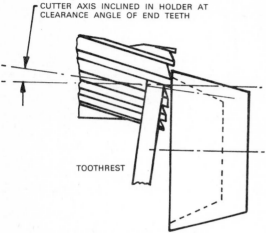

CUTTER AXIS INCLINED IN HOLDER AT CLEARANCE ANGLE OF END TEETH

TOOTHREST

Fig. 6.21. Regrinding clearance faces on an end mill.

Note that in all the above cases the teeth being ground are straight teeth, so the tooth rest holder is clamped to the machine table and moves with the cutter.

If a helical-toothed cutter is to be ground, the tooth rest must be fixed and not move with the cutter, in order to obtain the combination of rotational and linear movements required to follow the helix. The tooth rest is therefore attached to the wheel head and is not spring-loaded, but it must be wider than the wheel so that the cutter is fully controlled throughout its contact with the wheel. Due to its width, the tooth rest must be inclined to coincide with the helix angle of the wheel as shown in fig. 6.22. For indexing, the cutter is simply fed clear of the tooth rest and rotated until the next tooth can be located.

SETTING THE CUTTER TO PRODUCE THE CORRECT ANGLES

It is sometimes found that in order to set the cutter 'correctly' relative to the wheel, the grinder works by trial and error, adjusting the setting on successive teeth until the grinding marks register all over the existing face being ground. This is not good practice because if the cutter is already incorrect the error is perpetuated. Further, such a method can progressively increase errors in cutting angles. It is much better to calculate the setting required to produce the correct angle, irrespective of any errors which may already exist, and thus grind the cutters correctly.

GRINDING CLEARANCE FACES

The clearance face of a side-and-face cutter or a slitting saw can either be ground on the periphery of a disc wheel or on the face of a cup wheel.

A helical-toothed cutter is usually ground on the periphery of a wheel, in which case the clearance angle is obtained by raising the wheelhead so that the wheel centre is above the cutter centre.

The cutter and wheel centres are first aligned, a gauge often being provided with the machine for this purpose. The height of the tooth rest is then adjusted so that the cutting edge is at the same height as the cutter centre. The wheel head is then raised to give the correct clearance angle θ.

Referring to fig. 6.23

In triangle OAB

$$\sin \theta = \frac{OB}{OA} = \frac{h}{R_w}$$

and $\qquad h = R_w \sin \theta$

where h = height by which wheel centre is *raised* above cutter centre

R_w = radius of wheel

θ = clearance angle.

This method gives hollow-ground teeth and is not recommended except for helical slab mills, end mills, etc., which are difficult to grind in any other way.

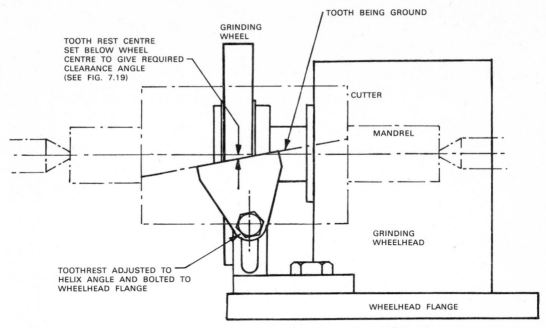

Fig. 6.22. Grinding clearance on a helical milling cutter. The cutter is 'ghosted' to allow details of the tooth rest to be shown.

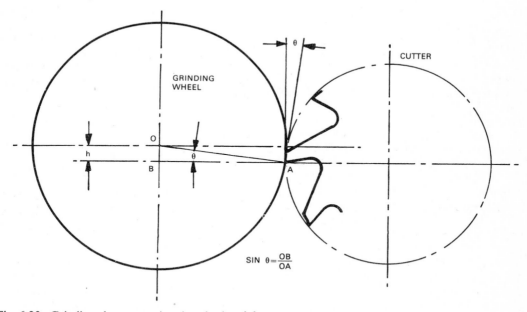

$$\text{SIN } \theta = \frac{OB}{OA}$$

Fig. 6.23. Grinding clearance using the wheel periphery.

To grind the clearance angle of a side-and-face cutter on a cup wheel as shown in fig. 6.20(a), the cutter is again set on the tooth rest so that the cutting edge is on the horizontal centre line of the cutter. The tooth edge is now lowered a distance h to give the required clearance angle.

From fig. 6.20(a):

In triangle OAB

$$\sin \theta = \frac{AB}{OA} = \frac{h}{R_c}$$

and $\qquad h = R_c \sin \theta$

where R_c = radius of cutter

h = distance by which cutting edge is *lowered* below cutter centre

θ = clearance angle.

In some cases the tooth rest holder has a screw adjustment, with graduations to enable the height adjustment to be made conveniently. If such equipment is not available the setting can be done with a height gauge.

GRINDING RAKE FACES

Sometimes a cutter requires grinding on the rake face. If a side-and-face cutter is ground too frequently on the clearance face the clearance land becomes too wide and can then be reduced by grinding the rake face. Also, continuous clearance grinding reduces the cutter diameter and hence the chip space, and it becomes necessary to regash the cutter to give adequate chip clearance. Certain cutters, notably form-relieved cutters, must be ground on the rake face only and the rake angle must be correct or the cutter will not produce the correct form.

To grind the rake face, a saucer-shaped wheel is used, as shown in fig. 6.21(b). The machine is set so that the wheel face is beyond the cutter centre line by an amount x to give the correct rake angle α. The cut is then set by lowering the tooth rest until the teeth contact the wheel face.

From fig. 6.20(b):

In triangle OAB

$$\sin \alpha = \frac{OB}{OA} = \frac{x}{R_c}$$

$$x = R_c \sin \alpha$$

where x = horizontal displacement of tooth edge when finished

R_c = cutter radius

α = rake angle.

Other multi-toothed cutters such as taps and reamers are ground in a similar manner and for all such work the following points are emphasised.

1. Light cuts only must be taken to avoid overheating and softening the tool or cutter.

2. Care must be taken to obtain the correct angles on the cutter to give an efficient cutting action.

3. Certain cutters, notably form-relieved cutters, reamers and taps, must be ground only on the rake face so that the form or diameter is not made incorrect.

4. To ensure that all teeth take an equal cutting load, the depth of grinding cut should not be changed until all teeth have been ground at the one setting.

5. In all this work and particularly in the case of small cutters, the operator's hand is close to the grinding wheel which, of necessity, may be unguarded. Concentration is necessary to avoid accidents.

QUESTIONS

1. A H.S.S. side and face cutter used in a particular operation is required to have a cutting speed of 20 metres/min and a feed of 0.1 mm/tooth. If the cutter is 150 mm diameter and has 10 teeth, calculate a suitable speed and table feed for the machine.

2. The cutter in question 1 is cutting a slot 15 mm deep in a block of metal whose overall length is 250 mm. Calculate the approach factor, the total length travelled and the cutting time for the job.

3. A face milling cutter has 8 teeth and is 100 mm diameter. It is cutting a block of mild steel 75 mm wide with a cutting speed of 120 metres/min and a feed rate of 0.05 mm/tooth. If the length of the block is 400 mm, calculate:
(a) spindle speed in rev/min;
(b) table feed in mm/min;
(c) cutting time.

4. Sketch three teeth of a side and face milling cutter and show clearly:
(a) Primary clearance.
(b) Secondary clearance.
(c) Rake angle.
(d) Chip clearance.

5. Sketch three teeth of a side and face milling cutter in plan view and show:
(a) Primary clearance.
(b) Secondary clearance.
(c) Rake angle.
(d) Chip clearance.
Give two reasons why the rate of metal removal with the side cutting teeth is limited and show how this can be increased with a staggered tooth side and face cutter.

6. When grinding some milling cutters the tooth rest travels with the cutter and for others it remains stationary. State the class of tooth which is ground with the tooth rest stationary and name two types of cutter in this class.

7. A side and face cutter of 150 mm diameter is being ground on its clearance face on a cup wheel. It is initially set with the cutting edge of the tooth at the same height as the cutter centre and the clearance angle is produced by lowering the tooth rest.
   Sketch the set-up and calculate the amount the cutting edge must be lowered to give a clearance angle of 7°.

8. An end mill is being ground on its peripheral teeth on the clearance face.
(a) Sketch the set up and explain why such cutters are ground using the periphery of a disc wheel rather than a cup wheel.
(b) The clearance angle is produced by initially setting the wheel and cutter centres at the same height and then raising the wheel head. If the grinding wheel is 200 mm diameter, by how much should the wheelhead be raised to give a clearance angle of 7°
(c) It is important in this case that the height of the work rest is correct. Explain how this is arranged when the work rest has to be inclined at the helix angle.

9. List the advantages of down-cut milling compared with up-cut milling and explain what special features a machine must have for it to be suitable for down-cut milling.

10. Form relieved cutters must be ground on the rake face and it is important, for other reasons than the cutting action, that the rake angle is maintained correct.
(a) Explain why this is so.
(b) Sketch the set up for grinding the rake face of a form relieved cutter and state what particular precautions must be taken by the operator in grinding such a cutter.

ANSWERS

1. 42 rev/min; 42 mm/min.

2. 45 mm; 295 mm; 6.55 min.

3. 382 rev/min; 191 mm/min; 2.18 min.

7. 9.14 mm.

7   Metal cutting with abrasive wheels

General Objective: *The student should be able to recognise and apply the principles of metal cutting to grinding, conforming to the appropriate regulations.*

INTRODUCTION

Grinding is essentially a true metal cutting process. That this is so is illustrated by fig. 7.1 which is a magnified photograph of the 'dust' obtained from grinding. The metallic particles are typical of metal cutting swarf, other materials present being the abrasive particles removed from the wheel.

Specific Objective: *The student should be able to name the features of grinding wheel construction, and the manner in which it achieves the self-sharpening effect.*

A grinding wheel has two constituents, the abrasive particles which do the cutting and the bond which holds the abrasive in the form of a wheel or disc. The abrasive particles project

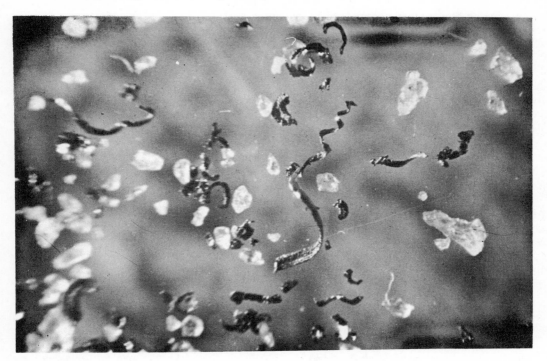

Fig. 7.1. Enlarged photograph of grinding dust. Note the similarity to the swarf produced by conventional cutting processes.

slightly from the disc and, being sharp-edged, have a clearance angle and negative rake. As the edge of the particle becomes dull the loads imposed upon it increase and it fails in one of two ways:

(a) It is torn out of the bond, thus disclosing other particles;

or (b) It is shattered and the fragments have sharp edges which continue cutting.

This is known as the *self-sharpening* action of a grinding wheel and if a grinding wheel is operating correctly it is inevitable that wheel wear occurs.

GRINDING FAULTS

Specific Objective: *The student should be able to describe the way in which a grinding wheel loses a self-sharpening effect.*

In the grinding process many chips of material are being removed simultaneously and ideally all of them should be thrown clear of the work. In practice some are retained in the gaps between the abrasive particles and clog or *load* the wheel. If the loading becomes excessive the abrasive particles do not protrude enough and the process changes from one of cutting to one of rubbing and excessive heat is generated.

If wheel loading occurs, a higher wheel speed and reduced feed may help, or changing the wheel to one with a softer[1] bond and finer grit size will improve matters.

The other major fault which can occur in the grinding process is *glazing*. Glazing occurs when the abrasive particles become blunt and lose their edges and the cutting load is not high enough to remove them from the bond. Again, the process changes from cutting to rubbing, with increased heat generation.

An obvious cure is to change to a softer (weaker bond) wheel, but the condition can be alleviated by increasing the amount of work

---

[1] 'Softer' in this context means a weaker bond. A 'soft' wheel has a weak bond and a 'hard' wheel a strong bond.

each grain has to do, thus increasing the load and improving the self-sharpening action. If the wheel speed is reduced or the work speed increased, the amount of metal removed per grain is increased and the likelihood of removing blunted grains is correspondingly increased, with a resulting improvement in action.

The opposite fault to glazing is excessive wheel wear, due to grains being removed before they become blunt. It follows that if the opposite remedial action is taken the work done by each grain is reduced. Thus, in this case, the wheel speed is increased and the work speed reduced.

Whichever fault occurs, and whichever remedy is chosen, the wheel should always be dressed by a diamond to restore its cutting action. No change in conditions will unclog a loaded wheel or re-form an excessively worn wheel.

SELECTION OF GRINDING WHEELS

Specific Objective: *The student should be able to describe the method of specification of grinding wheels and the influence of working conditions on the wheel selection.*

If a grinding wheel is correctly selected in the first place and its conditions of use are correct, none of the above faults should arise. The factors affecting the choice of a grinding wheel are:

(a) workpiece material
(b) angle of contact between work and wheel
(c) condition of machine
(d) wheel speed
(e) work speed.

The types of grinding wheel available must be considered in the light of these factors and the characteristics of the wheels must be balanced against the operating conditions. The characteristics of the wheel are indicated by the code on the cardboard pad on the side of the wheel. This follows a British Standard (BS 1814: 1952) and may be interpreted as follows:

A = Type of abrasive
60 = Grit size
N = Grade
8 = Structure
V = Type of bond

TYPE OF ABRASIVE

Two types of abrasive are generally used; aluminium oxide, signified by the letter A, and silicon carbide, signified by the letter C. Aluminium oxide is the slightly softer and should be used for grinding harder materials. Aluminium oxide wheels are usually white but may be red or brown in colour. Silicon carbide is a green material and wheels of this abrasive are known as green grit wheels.

GRIT SIZE

The abrasive grains used are accurately graded so that all grains in a wheel are of the same size. This is achieved by sieving or, with the finer grains, by centrifugal separation. The number indicates the number of meshes per unit length of the sieve they will just pass through, thus a large number indicates a fine grit and vice-versa. The grain sizes generally used are as follows:

Roughing work	8–24
Commercially-accepted finish	30–60
Fine finish	70–180
Very fine finish	220–600

GRADE OF WHEEL

This indicates the bond strength, a weak-bond wheel being known as a soft wheel and a strong-bond wheel as a hard one. A low letter of the alphabet indicates a soft wheel and a high letter a hard one. Thus the grading is as follows:

Very soft	A, B, C, D, E
Soft	F, G, H, I, J, K
Medium	L, M, N, O, P
Hard	Q, R, S, T, U
Very hard	V, W, X, Y, Z.

STRUCTURE

The structure of a wheel is an indication of the proportion of bond to abrasive. An open-structure wheel may have up to 30% bond while a close wheel may have only 10%. The structure is indicated by a simple numerical scale as follows:

Dense	0, 1, 2, 3	finishing
Medium	4, 5, 6, 7	general purpose
Open	8, 9, 10, 11, 12 etc.,	high metal removal rates.

TYPE OF BOND

The final letter indicates the type of bond used in the wheel, as follows:

Vitrified bond	V	Most used for general work
Resinoid bond	B	For fettling castings
Rubber bond	R	Strongest type. Used for thin cut-off wheels
Shellac bond	E	Thin wheels with a fine finish
Silicate bond	S	Mild action. Used for fine-edged tools. Very large wheels.

Thus, the wheel grade A 60 N 8 V above indicates an aluminium oxide (A) wheel of medium grit size (60), of medium hardness (N), having an open structure (8) and with a vitrified bond (V).

This information must now be balanced against the operating conditions. A good general rule is

harder work—softer wheel.

This applies to both bond strength and type of abrasive. If, therefore, we are grinding a hardened tool steel, an aluminium oxide wheel should be used. If a soft material is being ground, a silicon carbide wheel is generally used, simply to achieve the desired accuracy and finish. The harder grit retains its sharpness longer while the soft bond allows the relatively low loads to remove dull grains before glazing occurs.

ANGLE OF WHEEL CONTACT
This depends largely on the type of work being carried out and fig. 7.2 shows the variation in contact area between different machine set-ups. The rule is

large angle of contact—softer wheel.

MACHINE CONDITION
This refers generally to the rigidity of the machine and the amount of vibration present. The rule is

good rigid machine—softer wheel.

WHEEL SPEED
If the wheel speed is high, the work moves only a small distance while the abrasive grain passes. Thus at high wheel speeds each grain removes only a small piece of material and is therefore lightly loaded. The rule is

high wheel speed—softer wheel.

WORK SPEED
If the work speed is high, a grain of abrasive will be called upon to remove more material, each time it passes, than if the work speed is low. It follows that the rule is

high work speed—harder wheel.

Consider now a situation where a large inside diameter is being rough-ground in a hardened steel die ring. Because of the large diameter, the surface speed of the work will be relatively high and that of the wheel will be relatively low. The angle of contact will be large and we can decide from the known conditions that the type of grinding wheel required must be as follows:

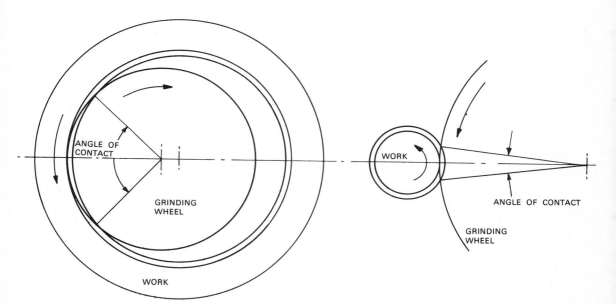

Fig. 7.2. Effect of type of work on angle of contact.

Factor	Conditions	Decision
(a) Wheel material	Hard work	Al_2O_3
(b) Grit size	Roughing	Coarse
(c) Wheel grade	High work speed	Hard
	Large contact angle	Soft
	Low wheel speed	Hard
	Old machine	Hard
(d) Wheel structure	Roughing	Open
(e) Type of bond	General	Vitrified bond

Thus we need an *Aluminium oxide* wheel of *coarse grit* with a *hard bond* of *open structure* and a *vitrified bond*. Such a wheel would be specified by

$$A \quad 20 \quad Q \quad 10 \quad V.$$

OPERATING FAULTS AND WORK CONDITIONS

Specific Objective: *The student should understand the causes of burning and cracking of work during grinding and how these problems may be overcome.*

In considering the cutting action of a grinding wheel, the self-sharpening action − or the lack of it − and wheel selection, we have been largely concerned with the effects of wheel selection and working conditions on the grinding wheel. In the end, however, the main consideration must be the effect of the operating conditions upon the workpiece, for the required end result is a workpiece made to the specification set out in the drawing. A glazed or loaded grinding wheel can be dressed and put into good operating condition, but a workpiece which is damaged by operating faults is scrap, and scrap is very expensive to produce but brings no income to the manufacturer.

Two operating faults which damage the workpiece are *burning* (or overheating) and *cracking*.

1. BURNING

One reason for grinding, rather than using 'conventional' cutting methods, is that the work is required to be hard when finished. If the grinding process itself causes local overheating, the work surface will become softened and thus damaged to the point of being scrapped.

It must be realised that the blue discolouration which occurs is the visible indication of burning and only the surface is oxidised or 'blued'. The softening will go deeper than this and it is not enough simply to take a light cut to skim off the discolouration. If, when burning occurs, the work is very close to its final size, it should be checked to ensure that its hardness still conforms to that specified.

Any machinist who has experienced overheating of the work when operating a lathe will appreciate that the causes of the trouble may be all or one of the following:

(a) the tool rubbing and not cutting;
(b) cutting speed too high;
(c) lack of coolant.

We have shown that grinding is fundamentally a cutting process and, this being so, the same things cause burning in grinding.

(a) *Wheel rubbing*

In the discussion on the self-sharpening action we have shown that grinding grits become blunt. If the self-sharpening action is not effective, the blunt grains will rub and cause burning. Similarly if the voids between the grits become clogged with chips removed from the work, the grits will no longer protrude enough to cut, and rubbing will take place. In both cases the 'tool' has become blunt and must be sharpened by dressing. If, after dressing, loading or glazing continue to cause burning, the wheel must be changed to eliminate the problem.

Similarly, in turning, if a tool is allowed to rub and not cut, due to not engaging the feed, overheating will occur. In the case of grinding, the 'feed' is the speed at which the work is fed past the wheel and, if the work speed is too *low*, rubbing and overheating occurs due to the wheel being too long in one place.

(b) *Cutting speed too high*

In any cutting operation the cutting speed is the velocity of the tool relative to the work. In grinding, the major component of the cutting speed is the peripheral speed of the wheel and if this is too great burning will occur.

(c) *Lack of coolant*

In conventional cutting, the function of the cutting fluid is to prevent overheating of the tool. In grinding, it is the work which must not be allowed to overheat. A copious flow of coolant should be directed at the work so that it is carried into the work zone, not kept out of it by the grinding wheel.

Thus burning can be avoided by:

  (i) not allowing the wheel to load or glaze;
  (ii) increasing the work speed;
  (iii) reducing the wheel speed;
  (iv) using plenty of coolant in the right direction.

2. CRACKING

Experience in other fields, notably heat treatment, should have shown the student that overheating followed by a 'drastic' quench causes cracking. Similarly, in grinding, if the work surface is allowed to overheat and the coolant is applied after the wheel has passed, surface cracks may be caused.

It must also be realised that heating during grinding is extremely local and does not penetrate very deeply into the body of the material. Thus a thin layer of material tries to expand, but is prevented from doing so by the mass of material beneath it. If the loads due to expansion produced in the surface layer are greater than its strength, then the surface layer will fail and cracks will be caused. Cracking can be prevented by:

  (a) using correct operating conditions to avoid overheating;
  (b) applying coolant correctly.

EFFECT OF WORKING CONDITIONS ON WHEEL WEAR

Specific Objective: *The student should be able to explain how the longitudinal feed rate can be adjusted to give optimum wheel wear during cylindrical grinding.*

If, during cylindrical grinding, the wheel width is much less than the length of the work, the process is carried out using the normal longitudinal feed of the machine. In this case the longitudinal rate of feed is important. If the feed per revolution of the work is one-third of the wheel width or less, the outer edges of the wheel will do all the cutting and wear will be concentrated at the edges, causing the wheel to wear to a convex profile. The centre portion of the wheel periphery, which does little cutting, rapidly becomes glazed. If, on the other hand, the work feed per revolution is about two-thirds of the wheel width, then on the rightward motion of the work the cutting will be done by the left-hand two-thirds of the wheel, and on the reverse stroke by the right-hand two-thirds. A little thought will show that these areas overlap and therefore the middle third of the wheel does twice as much cutting as the portions on either side. The wheel tends to wear to a concave profile, which is to be preferred. At the end of the stroke the wheel should not clear the work, but about one-third of its width should be allowed to run off.

If the work width is less than the wheel width, the work can be ground by the plunge-feed method. The wheel is carefully dressed and fed slowly and radially into the work, with no longitudinal feed in the usual sense. To reduce localised wheel-wear, the table should be reciprocated back and forth about 5 mm during cutting. This method, without the table oscillation, lends itself to form grinding.

WHEEL DRESSING

Specific Objective: *The student should be able to describe the dressing of grinding wheels using diamond trueing, crushing and star wheel dressing.*

Grinding wheels must be dressed for two reasons:

(a) to ensure that they cut freely and correctly;

(b) to impart to the wheel the correct geometric form for the machine or work for which it is to be used.

All methods of dressing restore correct cutting action to the wheel – they remove blunt grains of abrasive and the clogged surface of the wheel, thus revealing new sharp abrasive grits projecting from the bond, which can cut cleanly and efficiently. The method used then depends upon the machine and the type of operation to be carried out.

1. STAR WHEEL DRESSING

The star wheel dresser is a hand-held tool as shown in fig. 7.3. It consists of a handle at the end of which is mounted a number of 'star' wheels which, when pressed against a revolving grinding wheel, revolve with it and gouge out the bond. This releases the dulled grains and any wheel loading which has occurred, and allows the wheel to cut freely. Star dressers are normally only used on off-hand grinding machines where precision of form of the wheel is not necessary.

SAFETY NOTE

Star dressers should only be used on fairly substantial solid wheels where the load applied will not fracture the wheel. On smaller, more delicate wheels which have to be hand dressed, a small piece of an old wheel or similar hand abrasive material can be used and gently moved over the wheel surface. The author once worked for a spark-plug manufacturer – plug insulators are almost pure aluminium oxide. These insulators were ideal for the delicate hand dressing of the sides of small, fine, grinding wheels and almost everybody in the toolroom kept one in his toolbox for this purpose.

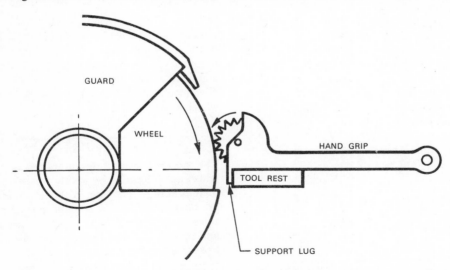

GUARD

WHEEL

HAND GRIP

TOOL REST

SUPPORT LUG

Fig. 7.3. Star wheel dresser used on an off-hand grinding machine. Note the lug against the front of the tool rest to give support and control of the wheel dresser.

175

2. DIAMOND DRESSING

When a wheel on a precision machine is being dressed using a diamond, the process is almost one of turning the wheel to the correct shape and form using the diamond as a tool. Thus, in the case of a surface grinder, the wheel periphery must be parallel to the table motion and dressing is carried out by mounting a diamond, in its holder, on the magnetic chuck, the longitudinal motion of the table being locked and the transverse table motion being used to traverse the diamond back and forth across the face of the wheel.

Note the position of the diamond, fig. 7.4, relative to the direction of rotation of the wheel. If for any reason the diamond becomes loose in its holder the rotation of the wheel will throw it clear, rather than dragging it under the wheel and creating a hazard.

If the wheel is to be used to grind the sides, as well as the base of a slot, the side of the wheel must be relieved as shown in fig. 7.5. This dressing can be carefully done by hand using a piece of abrasive, rather than a diamond. The abrasive stick or block, having a larger surface area, is less likely to dig in and damage the wheel.

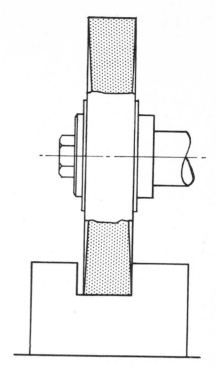

Fig. 7.5. Sides of wheel relieved for grinding groove faces.

If the wheel is to be used on a cylindrical grinder, then it must be dressed to the geometry of the machine. The diamond, in its holder, is mounted on the machine table and passed back and forth across the wheel, the cross feed being advanced a small amount at each pass until the wheel is clean and true.

SAFETY NOTE

The author has seen operators test a grinding wheel after dressing by gently touching it with their thumb — don't do it. A cut caused by a grinding wheel is also usually burned and the sides of the cut, being cauterised, are painful and take a long time to heal. Stop the machine to examine the wheel.

Where wheels are being dressed to a form, which is to be reproduced on the work, they can

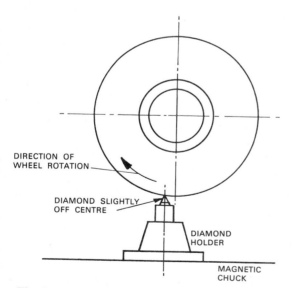

DIRECTION OF
WHEEL ROTATION

DIAMOND SLIGHTLY
OFF CENTRE

DIAMOND
HOLDER

MAGNETIC
CHUCK

Fig. 7.4. Wheel dressing on surface grinder.

be diamond dressed by causing the diamond to move in a path which generates the form required on the wheel. This is often done using an accurately made cam to control the position of the diamond as it passes across the wheel until the cam form is accurately reproduced on the wheel.

3. WHEEL CRUSHING

Grinding wheels used for form grinding can be accurately crushed to the correct form by forcing into the *slowly* rotating wheel a hardened steel roller which has itself been accurately ground to the correct form. A typical case is the forming of the wheel on a thread grinding machine for the production of accurate screw threads of high finish. It is important to note that the thread form is produced on the wheel as a series of parallel ribs, not as a screw thread.

The crushing roller is mounted on bearings on an axis parallel to the wheel axis, the whole assembly being mounted on a slideway which is hand fed into the grinding wheel. The grinding wheel is run at a very slow speed and the crushing roller fed, by hand, gradually into the grinding wheel until the form is reproduced on the wheel.

When the wheel is formed ready for grinding, the wheel head is turned through the helix angle of the thread to be produced, and, with the wheel running at the correct speed for grinding, the work rotates slowly and moves axially one pitch of the thread per revolution of the work. The wheel is fed in to the full depth of the thread to be produced and grinding continues until the thread is completed. In feed, traverse along the thread and withdrawal are automatically controlled by stops.

Thread grinding is by no means the only type of work for which wheels are crushed to form but it is a common use for this process. Other forms can be produced and in these cases, where no helix is involved, the wheel and work axes are parallel and there is no axial motion, a plunge feed at right angles to the work axis being the only feed action involved.

SAFETY IN GRINDING

Specific Objective: *The student should be aware of the hazards in using abrasive wheels and be familiar with the abrasive wheel regulations.*

By the very nature of the process — an abrasive wheel rotating at high peripheral speeds to remove large numbers of small particles from the work, and in so doing losing some of its own abrasive grains — grinding carries with it two basic hazards: **eye damage from flying particles and more physical damage from flying chunks of burst wheels**. Although both sound dramatic, both can be avoided by the sensible application of safety rules.

EYE HAZARDS

All grinding areas should be regarded as eye hazard areas and safety glasses should be worn at all times by everybody in the area — not just the operator. At the same time wheels should only be run and used with the guards correctly positioned and used. The guards not only direct flying fragments of a burst wheel into safe orbits but they also direct sparks and dust into areas where they can do least harm.

WHEEL BREAKAGES

Grinding wheels burst in use very rarely. In a lifetime in engineering the author has seen two cases, both of which would have been avoided had the safety rules been followed. Grinding wheel manufacturers take all possible precautions against supplying faulty wheels and it is rare for a faulty wheel to leave the manufacturers' despatch department. Catastrophic failure of a grinding wheel is usually due to one or more of the following factors:

1. *Damage in transport*

All abrasive wheels are fragile and must be handled carefully to avoid damage during

177

transport. The Abrasive Industries Association makes the following recommendations:

(a) Wheels must always be handled carefully to avoid dropping or bumping.

(b) Wheels should not be rolled. If large wheels which are too heavy to be carried must be rolled, a soft resilient floor surface should be provided.

(c) Wheels too heavy to be carried should be moved in suitable trucks or conveyances which provide proper support.

(d) During transport, wheels should be stacked carefully on trucks in a stable manner to prevent them toppling over. Castings, tools and other equipment should never be transported with grinding wheels.

2. *Damage in storage*
Special racks and bins should be used for storing grinding wheels. The wheels should never be kept in a drawer or on a shelf with other equipment.

Plain and tapered wheels should be stored on edge in a cradle to prevent them rolling out of the rack. Thin wheels should be laid flat on a horizontal surface to prevent warping and large cylinder and cup wheels may also be stored on their flat surfaces.

All wheels which are stacked one on top of the other should be separated by sheets of corrugated cardboard to prevent one wheel damaging another when being placed into or withdrawn from the rack.

3. *Excessive speed*
Grinding wheels burst because the centrifugal force set up by the rotational motion exceeds the strength of the wheel material. The stress in a rotating disc is proportional to the *square* of the peripheral speed. Thus, from a safety standpoint, wheel speeds are critical and wheels are marked with their maximum safe speed. At the same time all grinding wheel spindles must be marked with the spindle speed and under no circumstances must the specified speed for a new wheel be exceeded.

As the diameter of the wheel is reduced by successive dressings, the speed in rev/min may be increased as long as the peripheral speed recommended is not exceeded. Thus if a wheel of initial diameter D_1 mm has a recommended speed in rev/min of N_1 then:

$$\text{Peripheral speed in m/min} = \frac{\pi D_1 N_1}{1000} \text{ m/min}$$

If the wheel is now dressed down to a diameter of D_2 mm, its new allowable speed in rev/min will be N_2 and therefore,

$$\text{Peripheral speed in m/min} = \frac{\pi D_2 N_2}{1000} \text{ m/min}$$

As these peripheral speeds are equal in that the initial peripheral speed must not be exceeded then,

$$\frac{\pi D_2 N_2}{1000} = \frac{\pi D_1 N_1}{1000}$$

and

$$N_2 = \frac{D_1 N_1}{D_2}$$

4. *Incorrect mounting*
Before an abrasive wheel is mounted there are certain precautions which should be taken to ensure that, in spite of all the precautions mentioned earlier, the wheel has survived the rigours of transport and storage and arrived at the machine undamaged. The wheel should be carefully brushed and examined for cracks and defects.

If a wheel is cracked, its strength is reduced and it will burst at a much lower speed than that recommended by the manufacturer. Cracks may not be visible to the naked eye but they can be detected by 'ringing' the wheel. If a sound wheel is supported by a finger in its centre hole and tapped lightly with a non-metallic material – a pencil will do – a pure ringing sound will result. If the wheel is cracked it will sound flat and dead. This can be easily demonstrated in a kitchen with two cups—a cracked and a sound one. If the cups are placed side by side on a table and tapped with a pencil the difference is immediately apparent.

Having ensured that the wheel is not cracked it may be mounted on the wheel spindle. The method of mounting varies depending on the type of wheel and spindle but, in general, the following rules should be followed:

(a) The wheel should never be mounted on a machine for which it is not intended.

(b) The wheel should fit easily, but not loosely, onto the spindle.

(c) The direction of *tightening* the nut should be *opposite* to the direction of wheel direction.

(d) The correct flanges (and washers if used) must be used to mount the wheel. Improvised bits and pieces are **dangerous**.

(e) Paper washers, or blotters, should always be used except on tapered wheels. These washers are not just to provide information about the wheel specification—their main purpose is to evenly distribute the clamping force applied by the flanges.

(f) Flanges should be of mild steel or a material of equal or greater strength and rigidity. Cast iron should never be used. The two flanges should be of equal outside diameter and diameter of recess to give identical bearing areas on either side of the wheel. The clamping surfaces should be true and there should be no rough edges or protrusions.

(g) The flanges should be assembled correctly, the inner flange being keyed to the spindle, and the nut assembled only tightly enough to secure the wheel. If the recommended tightening torque is known a torque wrench should be used.

(h) Before starting the wheel, the guard must be correctly and securely positioned.

Fig. 7.6 shows a selection of different grinding wheels correctly mounted. Note that the construction of the assembly conforms to the rules listed, paper washers being used in all cases except for the tapered wheels, flanges clamping identical areas on both sides of the wheel, and the flanges being recessed in order to clear the edge of the wheel hole.

BALANCING OF GRINDING WHEELS

Specific Objective: *The student should be able to demonstrate the balancing of a grinding wheel.*

If any mass rotating at high speed is out-of-balance, vibrations occur which, in most cases, can cause damage to the machinery of which the mass is part. In the case of grinding wheels, the damage may be to the wheel, the spindle and bearings, the work, and in catastrophic cases, the operator.

Grinding wheels are balanced by the makers before leaving the factory, but for precision grinding, closer limits of balance than those provided by the makers may be necessary. Three stages are necessary in the balancing of a grinding wheel:

1. Determine the position for the balancing mass required.
2. Add balancing masses.
3. Check the balance of the wheel assembly.

To determine the position of the required balancing mass, the wheel may be mounted on a balancing arbor and placed on a balancing stand. This consists of a pair of knife edges set parallel and accurately horizontal. A wheel which is correctly balanced will remain stationary on the stand whatever its initial position. If the wheel is out of balance it will roll on the stand and oscillate, only becoming stationary when the out of balance mass is at the bottom. If the wheel remains stationary when first placed on the stand it should be turned through 90° just in case the out-of-balance mass was, by chance, placed in the lowest position.

Having allowed the wheel to come to rest, it should be marked to indicate the position where the balancing mass is to be added or removed. Balance may be achieved by the addition of heavy paint or the removal of lead from the wheel bush. In the case of larger wheels, the flanges are provided with sliding weights which can be adjusted in slots in the flanges until balance is achieved.

179

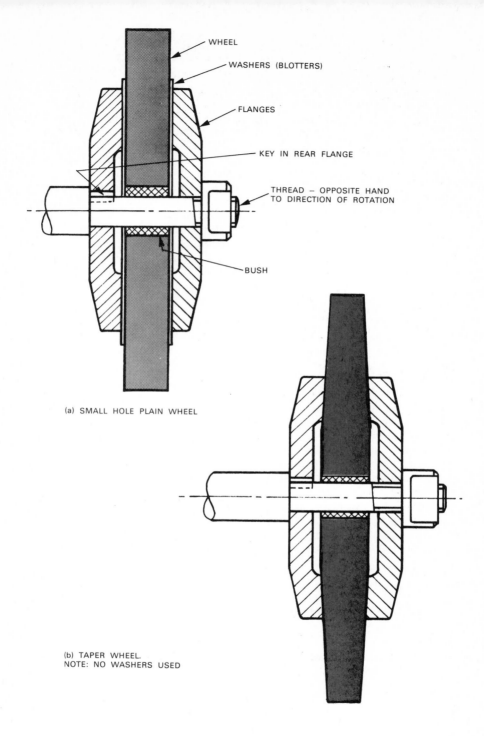

WHEEL

WASHERS (BLOTTERS)

FLANGES

KEY IN REAR FLANGE

THREAD — OPPOSITE HAND
TO DIRECTION OF ROTATION

BUSH

(a) SMALL HOLE PLAIN WHEEL

(b) TAPER WHEEL.
NOTE: NO WASHERS USED

180

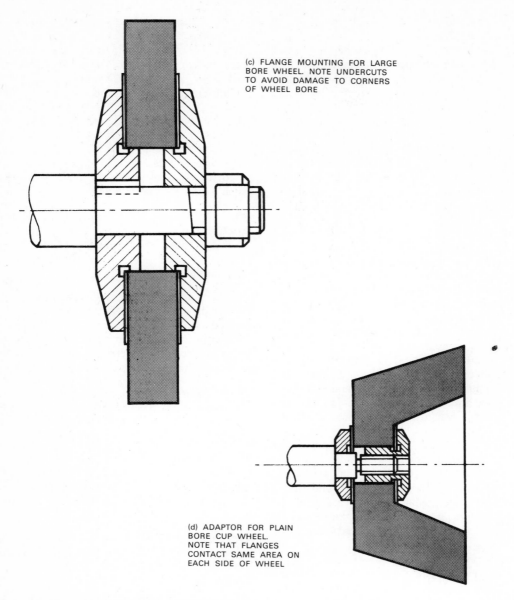

(c) FLANGE MOUNTING FOR LARGE BORE WHEEL. NOTE UNDERCUTS TO AVOID DAMAGE TO CORNERS OF WHEEL BORE

(d) ADAPTOR FOR PLAIN BORE CUP WHEEL. NOTE THAT FLANGES CONTACT SAME AREA ON EACH SIDE OF WHEEL

Fig. 7.6. Grinding wheels correctly mounted.

The method described is very much one of trial-and-error, but it is perfectly satisfactory. A more modern method, using an *accelerometer*, utilises a probe attached to the machine which displays the balancing mass required on a meter, and a stroboscope which enables the position of the weight to be determined. Students are probably familiar with similar equipment used to balance the wheels of motor cars. As with motor cars, this equipment enables the wheel to be balanced on the machine and any out-of-balance in the spindle is also corrected.

THE ABRASIVE WHEEL REGULATIONS

In 1970 a Statutory Instrument of Government was laid before Parliament, which came into operation in April 1971 and is known as *The Abrasive Wheel Regulations 1970*. As with most legal documents the instrument is drawn up in language which, although confusing to the layman, is necessary to avoid ambiguities. The first five regulations deal with the date of introduction, interpretation and definitions used, application and operation, and exceptions and exemptions from the regulations. Regulations which may be termed 'Technical' are as follows:

No. 6: Speeds of abrasive wheels

This requires that the wheel shall be clearly marked with its maximum speed, and that it shall not be operated at a speed in excess of that specified; but, it states that the speed may be increased as its diameter is reduced (see p. 178).

No. 7: Speeds of spindles

This regulation requires that each spindle shall be clearly marked with its maximum speed (or speeds) and that air-driven spindles shall be governed so that the stated maximum speed is not exceeded.

No. 8: Mounting

This simply states that: 'Every abrasive wheel shall be properly mounted.'

No. 9: Training and appointment of persons to mount abrasive wheels

This is most important in that it requires that *no person shall mount an abrasive wheel unless he has been trained · or is competent to do so.* Additionally a register must be kept by the employer naming the person appointed and the class of wheel he is appointed to mount.

No. 10: Provision of Guards

This requires that a guard shall be provided and used on all abrasive wheels unless the use of such a guard is impracticable.

No. 11: Construction and maintenance of guards

Guards must be properly constructed, maintained and enclose the wheel so as to contain the wheel if it bursts.

No. 12: Tapered wheels and protection flanges

Where the exposed arc of the wheel in operation exceeds 180°, the wheel must be tapered from its centre to its periphery and suitable flanges must be used. Diameters and the construction of flanges for all wheels are specified.

No. 13: Selection of abrasive wheels

The wheel must be correctly selected, not only to do the job for which it is required, but to minimise the danger to the operator.

No. 14: Machine controls

The controls for starting and stopping the machine shall not only be efficient but must be placed conveniently for the operator.

No. 15: Rests

Any work rest must be securely fixed and as close as possible to the wheel; it must be substantially constructed and properly maintained.

The remaining regulations, numbers 16 to 19 inclusive, deal with cautionary notices, condition of floors, duties of employed persons and sale or hire of machinery. Additionally, there is a schedule to regulation 9 detailing the particulars of training of persons approved to mount abrasive wheels.

Much of the previous work in this chapter has dealt with the requirements of these regulations which have been outlined briefly above. Additionally the author recommends to the student the following literature published by Her Majesty's Stationery Office for the Department of Employment.

    1. Health and Safety at Work booklet No. 4: *Safety in the Use of Abrasive Wheels.*
    2. Training Advisory Leaflet No. 2: *Advice on the Mounting of Abrasive Wheels.*
    3. Statutory Instrument No. 535 (1970): *The Abrasive Wheels Regulations 1970.*

Much of this chapter has been concerned with safety. Many of the regulations are recognisable as practices which have long been in use by those concerned with grinding operations. The Abrasive Wheels Regulations incorporate these into a document which gives the backing of the Law to safe practices.

QUESTIONS

1. A grinding wheel contains two constituents: the abrasive and the bond. List two types of abrasive and state an application of each. List four types of bond and state an application of each.

2. A grinding wheel is designated:
      A 40 R 7 V
Explain the meaning of the terms in this code and name the purpose for which it may be used.

3. A grinding wheel is required for hollow grinding the edge of bread knives. Specify a suitable wheel using the BSS code for the designation of grinding wheels. *Note:* Requirements are a fine finish with a mild cutting action.

4. Certain bronze bushes are finished in the bores by grinding. Explain why this may be done when bronze can be easily cut by conventional methods. Select a suitable wheel for such a process.

5. In a grinding operation there are certain factors which cannot readily be changed; they are:
    (a) Work material.
    (b) Finish required.
    (c) Arc of contact.
    (d) Type of grinding machine.
Explain how each influences the choice of wheel to be used.

6. In a grinding operation there are certain factors which can be more easily changed than those considered in question 5; they are:
    (a) Work speed.
    (b) Wheel speed.
    (c) Machine condition.
    (d) Skill of operator.
Explain how each influences the choice of wheel to be used.

7. Explain what is meant by the term 'self-sharpening effect'.
    Two common operating faults in grinding are loading and glazing. Explain what is meant by each term, how the fault is caused and what action may be taken to cure it.

8. In order to guard against grinding wheels bursting certain precautions should be taken before the wheel is mounted, during mounting and after it is mounted. List these precautions.

9. A grinding wheel, 150 mm diameter, is marked with its maximum safe speed of 4000 rev/min. To what diameter must it be dressed-down before the speed can be increased, if the next highest speed on the machine is 6000 rev/min.

ANSWER

9. 100 mm.

Appendix 1

Selected ISO Fits – Hole Basis

EXTRACTED FROM BS 4500[1]

Reproduced on pages 186–7

The ISO system provides a great many hole and shaft tolerances so as to cater for a very wide range of conditions. However, experience shows that the majority of fit conditions required for normal engineering products can be provided by a quite limited selection of tolerances.

The following selected hole and shaft tolerances have been found to be commonly applied:

Selected hole tolerances: **H7; H8; H9; H11.**

Selected shaft tolerances: **c11; d10; e9; f7; g6; h6; k6; n6; p6; s6**

The table in this data sheet shows a range of fits derived from these selected hole and shaft tolerances. As will be seen, it covers fits from loose clearance to heavy interference and it may therefore be found to be suitable for most normal requirements. Many users may in fact find that their needs are met by a further selection within this selected range.

It should be noted, however, that this table is offered only as an example of how a restricted selection of fits can be made. It is clearly impossible to recommend selections of fits which are appropriate to all sections of industry, but it must be emphasised that a user who decides upon a selected range will always

[1] This extract from BS 4500: 1969: *Limits and Fits for Engineering* is reproduced by permission of the British Standards Institution, 2 Park Street, London W1A 2BS, from whom copies of the complete standard may be obtained.

enjoy the economic advantages this conveys. Once he has installed the necessary tooling and gauging facilities, he can combine his selected hole and shaft tolerances in different ways without any additional investment in tools and equipment.

For example, if it is assumed that the range of fits shown in the table has been adopted but that, for a particular application the fit **H8–f7** is appropriate but provides rather too much variation, the hole tolerance **H7** could equally well be associated with the shaft **f7** and may provide exactly what is required without necessitating any additional tooling.

For most general applications it is usual to recommend hole basis fits as, except in the realm of very large sizes where the effects of temperature play a large part, it is usually considered easier to manufacture and measure the male member of a fit and it is thus desirable to be able to allocate the larger part of the tolerance available to the hole and adjust the shaft to suit.

In some circumstances, however, it may in fact be preferable to employ a shaft-basis. For example, in the case of driving shafts where a single shaft may have to accommodate a variety of accessories such as couplings, bearings, collars etc., it is preferable to maintain a constant diameter for the permanent member, which is the shaft, and vary the bore of the accessories. For use in applications of this kind, a selection of shaft basis fits is provided in Data Sheet 4500B.

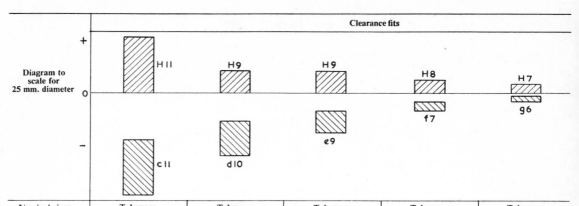

Clearance fits — Diagram to scale for 25 mm. diameter (H11, c11, H9, d10, H9, e9, H8, f7, H7, g6)

Nominal sizes		Tolerance		Tolerance		Tolerance		Tolerance		Tolerance	
Over	To	H11	c11	H9	d10	H9	e9	H8	f7	H7	g6
mm	mm	0·001 mm	0·001 mm	0·001 mm	0·001 mm	0·001 mm	0·001 mm	0·001 mm	0·001 mm	0·001 mm	0·001 mm
—	3	+ 60 / 0	− 60 / − 120	+ 25 / 0	− 20 / − 60	+ 25 / 0	− 14 / − 39	+ 14 / 0	− 6 / − 16	+ 10 / 0	− 2 / − 8
3	6	+ 75 / 0	− 70 / − 145	+ 30 / 0	− 30 / − 78	+ 30 / 0	− 20 / − 50	+ 18 / 0	− 10 / − 22	+ 12 / 0	− 4 / − 12
6	10	+ 90 / 0	− 80 / − 170	+ 36 / 0	− 40 / − 98	+ 36 / 0	− 25 / − 61	+ 22 / 0	− 13 / − 28	+ 15 / 0	− 5 / − 14
10	18	+ 110 / 0	− 95 / − 205	+ 43 / 0	− 50 / − 120	+ 43 / 0	− 32 / − 75	+ 27 / 0	− 16 / − 34	+ 18 / 0	− 6 / − 17
18	30	+ 130 / 0	− 110 / − 240	+ 52 / 0	− 65 / − 149	+ 52 / 0	− 40 / − 92	+ 33 / 0	− 20 / − 41	+ 21 / 0	− 7 / − 20
30	40	+ 160 / 0	− 120 / − 280	+ 62 / 0	− 80 / − 180	+ 62 / 0	− 50 / − 112	+ 39 / 0	− 25 / − 50	+ 25 / 0	− 9 / − 25
40	50	+ 160 / 0	− 130 / − 290								
50	65	+ 190 / 0	− 140 / − 330	+ 74 / 0	− 100 / − 220	+ 74 / 0	− 60 / − 134	+ 46 / 0	− 30 / − 60	+ 30 / 0	− 10 / − 29
65	80	+ 190 / 0	− 150 / − 340								
80	100	+ 220 / 0	− 170 / − 390	+ 87 / 0	− 120 / − 260	+ 87 / 0	− 72 / − 159	+ 54 / 0	− 36 / − 71	+ 35 / 0	− 12 / − 34
100	120	+ 220 / 0	− 180 / − 400								
120	140	+ 250 / 0	− 200 / − 450	+ 100 / 0	− 145 / − 305	+ 100 / 0	− 84 / − 185	+ 63 / 0	− 43 / − 83	+ 40 / 0	− 14 / − 39
140	160	+ 250 / 0	− 210 / − 460								
160	180	+ 250 / 0	− 230 / − 480								
180	200	+ 290 / 0	− 240 / − 530	+ 115 / 0	− 170 / − 355	+ 115 / 0	− 100 / − 215	+ 72 / 0	− 50 / − 96	+ 46 / 0	− 15 / − 44
200	225	+ 290 / 0	− 260 / − 550								
225	250	+ 290 / 0	− 280 / − 570								
250	280	+ 320 / 0	− 300 / − 620	+ 130 / 0	− 190 / − 400	+ 130 / 0	− 110 / − 240	+ 81 / 0	− 56 / − 108	+ 52 / 0	− 17 / − 49
280	315	+ 320 / 0	− 330 / − 650								
315	355	+ 360 / 0	− 360 / − 720	+ 140 / 0	− 210 / − 440	+ 140 / 0	− 125 / − 265	+ 89 / 0	− 62 / − 119	+ 57 / 0	− 18 / − 54
355	400	+ 360 / 0	− 400 / − 760								
400	450	+ 400 / 0	− 440 / − 840	+ 155 / 0	− 230 / − 480	+ 155 / 0	− 135 / − 290	+ 97 / 0	− 68 / − 131	+ 63 / 0	− 20 / − 60
450	500	+ 400 / 0	− 480 / − 880								

DATA SHEET

	Transition fits			Interference fits	

H7 / h6   **H7 / k6**   **H7 / n6**   **H7 / p6**   **H7 / s6**

Holes   Shafts

Tolerance		Tolerance		Tolerance		Tolerance		Tolerance		Nominal sizes	
H7	h6	H7	k6	H7	n6	H7	p6	H7	s6	Over	To
0·001 mm	0·001 mm	0·001 mm	0·001 mm	0·001 mm	0·001 mm	0·001 mm	0·001 mm	0·001 mm	0·001 mm	mm	mm
+10 / 0	−6 / 0	+10 / 0	+6 / +0	+10 / 0	+10 / +4	+10 / 0	+12 / +6	+10 / 0	+20 / +14	—	3
+12 / 0	−8 / 0	+12 / 0	+9 / +1	+12 / 0	+16 / +8	+12 / 0	+20 / +12	+12 / 0	+27 / +19	3	6
+15 / 0	−9 / 0	+15 / 0	+10 / +1	+15 / 0	+19 / +10	+15 / 0	+24 / +15	+15 / 0	+32 / +23	6	10
+18 / 0	−11 / 0	+18 / 0	+12 / +1	+18 / 0	+23 / +12	+18 / 0	+29 / +18	+18 / 0	+39 / +28	10	18
+21 / 0	−13 / 0	+21 / 0	+15 / +2	+21 / 0	+28 / +15	+21 / 0	+35 / +22	+21 / 0	+48 / +35	18	30
+25 / 0	−16 / 0	+25 / 0	+18 / +2	+25 / 0	+33 / +17	+25 / 0	+42 / +26	+25 / 0	+59 / +43	30	40
										40	50
+30 / 0	−19 / 0	+30 / 0	+21 / +2	+30 / 0	+39 / +20	+30 / 0	+51 / +32	+30 / 0	+72 / +53	50	65
								+30 / 0	+78 / +59	65	80
+35 / 0	−22 / 0	+35 / 0	+25 / +3	+35 / 0	+45 / +23	+35 / 0	+59 / +37	+35 / 0	+93 / +71	80	100
								+35 / 0	+101 / +79	100	120
+40 / 0	−25 / 0	+40 / 0	+28 / +3	+40 / 0	+52 / +27	+40 / 0	+68 / +43	+40 / 0	+117 / +92	120	140
								+40 / 0	+125 / +100	140	160
								+40 / 0	+133 / +108	160	180
+46 / 0	−29 / 0	+46 / 0	+33 / +4	+46 / 0	+60 / +31	+46 / 0	+79 / +50	+46 / 0	+151 / +122	180	200
								+46 / 0	+159 / +130	200	225
								+46 / 0	+169 / +140	225	250
+52 / 0	−32 / 0	+52 / 0	+36 / +4	+52 / 0	+66 / +34	+52 / 0	+88 / +56	+52 / 0	+190 / +158	250	280
								+52 / 0	+202 / +170	280	315
+57 / 0	−36 / 0	+57 / 0	+40 / +4	+57 / 0	+73 / +37	+57 / 0	+98 / +62	+57 / 0	+226 / +190	315	355
								+57 / 0	+244 / +208	355	400
+63 / 0	−40 / 0	+63 / 0	+45 / +5	+63 / 0	+80 / +40	+63 / 0	+108 / +68	+63 / 0	+272 / +232	400	450
								+63 / 0	+292 / +252	450	500

B.S. 4500 A

Appendix 2

Selected ISO Fits – Shaft Basis

EXTRACTED FROM BS 4500[1]

Reproduced on pages 190–1

The ISO system provides a great many hole and shaft tolerances so as to cater for a very wide range of conditions. However, experience shows that the majority of fit conditions required for normal engineering products can be provided by a quite limited selection of tolerances.

The following selected hole and shaft tolerances have been found to be commonly applied:

Selected hole tolerances: **H7**; **H8**; **H9**; **H11**

Selected shaft tolerances: **c11**; **d10**; **e9**; **f7**; **g6**; **h6**; **k6**; **n6**; **p6**; **s6**

For most general applications it is usual to recommend hole basis fits, i.e., fits in which the design size for the hole is the basic size and variations in the grade of fit for any particular hole are obtained by varying the clearance and the tolerance on the shaft. Data Sheet 4500A gives a range of hole basis fits derived from the selected hole and shaft tolerances above.

In some circumstances, however, it may in fact be preferable to employ a shaft basis. For example, in the case of driving shafts where a single shaft may have to accommodate a variety of accessories such as couplings, bearings, collars etc., it is preferable to maintain a constant diameter for the permanent member, which is the shaft, and vary the bore of the

[1] This extract from BS 4500: 1969: *Limits and Fits for Engineering* is reproduced by permission of the British Standards Institution, 2 Park Street, London W1A 2BS, from whom copies of the complete standard may be obtained.

accessories. Shaft basis fits also provide a useful economy where bar stock material is available to standard shaft tolerances of the ISO System.

For the benefit of those wishing to use shaft basis fits, this Data Sheet shows the shaft basis equivalents of the hole basis fits in Data Sheet 4500A. They are all direct conversions except that the fit **H9–d10**, instead of being converted to **D9–h10** is adjusted to **D10–h9** to avoid introducing the additional shaft tolerance **h10**.

As will be seen, the table covers fits from loose clearance to heavy interference and may therefore be found suitable for most normal requirements. Many users may in fact find that their needs are met by a further selection within this selected range.

It should be noted, however, that this Table is offered only as an example of how a restricted selection of fits can be made. It is clearly impossible to recommend selections of fits which are appropriate to all sections of industry, but it must be emphasised that a user who decides upon a selected range will always enjoy the economic advantages this conveys. Once he has installed the necessary tooling and gauging facilities, he can combine his selected hole and shaft tolerances in different ways without any additional investment in tools and equipment.

For example, if it is assumed that the range of fits shown in the table has been adopted but that, for a particular application the fit **h7–F8** is appropriate but provides rather too much variation, the shaft tolerance **h6** could equally well be associated with the hole **F8** and may provide exactly what is required without necessitating any additional tooling.

Diagram to scale for 25 mm diameter

C11   D10   E9   F8   G7

h11   h9   h9   h7   h6

Nominal sizes		Tolerance		Tolerance		Tolerance		Tolerance		Tolerance	
Over	To	h11	C11	h9	D10	h9	E9	h7	F8	h6	G7
mm	mm	0·001 mm	0·001 mm	0·001 mm	0·001 mm	0·001 mm	0·001 mm	0·001 mm	0·001 mm	0·001 mm	0·001 mm
—	3	0 / −60	+120 / +60	0 / −25	+60 / +20	0 / −25	+39 / +14	0 / −10	+20 / +6	0 / −6	+12 / +2
3	6	0 / −75	+145 / +70	0 / −30	+78 / +30	0 / −30	+50 / +20	0 / −12	+28 / +10	0 / −8	+16 / +4
6	10	0 / −90	+170 / +80	0 / −36	+98 / +40	0 / −36	+61 / +25	0 / −15	+35 / +13	0 / −9	+20 / +5
10	18	0 / −110	+205 / +95	0 / −43	+120 / +50	0 / −43	+75 / +32	0 / −18	+43 / +16	0 / −11	+24 / +6
18	30	0 / −130	+240 / +110	0 / −52	+149 / +65	0 / −52	+92 / +40	0 / −21	+53 / +20	0 / −13	+28 / +7
30	40	0 / −160	+280 / +120	0 / −62	+180 / +80	0 / −62	+112 / +50	0 / −25	+64 / +25	0 / −16	+34 / +9
40	50	0 / −160	+290 / +130								
50	65	0 / −190	+330 / +140	0 / −74	+220 / +100	0 / −74	+134 / +60	0 / −30	+76 / +30	0 / −19	+40 / +10
65	80	0 / −190	+340 / +150								
80	100	0 / −220	+390 / +170	0 / −87	+260 / +120	0 / −87	+159 / +72	0 / −35	+90 / +36	0 / −22	+47 / +12
100	120	0 / −220	+400 / +180								
120	140	0 / −250	+450 / +200	0 / −100	+305 / +145	0 / −100	+185 / +85	0 / −40	+106 / +43	0 / −25	+54 / +14
140	160	0 / −250	+460 / +210								
160	180	0 / −250	+480 / +230								
180	200	0 / −290	+530 / +240	0 / −115	+355 / +170	0 / −115	+215 / +100	0 / −46	+122 / +50	0 / −29	+61 / +15
200	225	0 / −290	+550 / +260								
225	250	0 / −290	+570 / +280								
250	280	0 / −320	+620 / +300	0 / −130	+400 / +190	0 / −130	+240 / +110	0 / −52	+137 / +56	0 / −32	+62 / +17
280	315	0 / −320	+650 / +330								
315	355	0 / −360	+720 / +360	0 / −140	+440 / +210	0 / −140	+265 / +125	0 / −57	+151 / +62	0 / −36	+75 / +18
355	400	0 / −360	+760 / +400								
400	450	0 / −400	+840 / +440	0 / −155	+480 / +230	0 / −155	+290 / +135	0 / −63	+165 / +68	0 / −40	+83 / +20
450	500	0 / −400	+880 / +480								

DATA SHEET

	Transition fits			Interference fits			Holes / Shafts

Diagram (upper): H7 / h6 — K7 / h6 — N7 / h6 — P7 / h6 — S7 / h6 ; Holes, Shafts

Tolerance		Tolerance		Tolerance		Tolerance		Tolerance		Nominal sizes	
h6	H7	h6	K7	h6	N7	h6	P7	h6	S7	Over	To
0·001 mm	0·001 mm	0·001 mm	0·001 mm	0·001 mm	0·001 mm	0·001 mm	0·001 mm	0·001 mm	0·001 mm	mm	mm
0 / −6	+10 / 0	0 / −6	0 / −10	0 / −6	−4 / −14	0 / −6	−6 / −16	0 / −6	−14 / −24	—	3
0 / −8	+12 / 0	0 / −8	+3 / −9	0 / −8	−4 / −16	0 / −8	−8 / −20	0 / −8	−15 / −27	3	6
0 / −9	+15 / 0	0 / −9	+5 / −10	0 / −9	−4 / −19	0 / −9	−9 / −24	0 / −9	−17 / −32	6	10
0 / −11	+18 / 0	0 / −11	+6 / −12	0 / −11	−5 / −23	0 / −11	−11 / −29	0 / −11	−21 / −39	10	18
0 / −13	+21 / 0	0 / −13	+6 / −15	0 / −13	−7 / −28	0 / −13	−14 / −35	0 / −13	−27 / −48	18	30
0 / −16	+25 / 0	0 / −16	+7 / −18	0 / −16	−8 / −33	0 / −16	−17 / −42	0 / −16	−34 / −59	30	40
0 / −16	+25 / 0	0 / −16	+7 / −18	0 / −16	−8 / −33	0 / −16	−17 / −42	0 / −16	−34 / −59	40	50
0 / −19	+30 / 0	0 / −19	+9 / −21	0 / −19	−9 / −39	0 / −19	−21 / −51	0 / −19	−42 / −72	50	65
0 / −19	+30 / 0	0 / −19	+9 / −21	0 / −19	−9 / −39	0 / −19	−21 / −51	0 / −19	−48 / −78	65	80
0 / −22	+35 / 0	0 / −22	+10 / −25	0 / −22	−10 / −45	0 / −22	−24 / −59	0 / −22	−58 / −93	80	100
0 / −22	+35 / 0	0 / −22	+10 / −25	0 / −22	−10 / −45	0 / −22	−24 / −59	0 / −22	−66 / −101	100	120
0 / −25	+40 / 0	0 / −25	+12 / −28	0 / −25	−12 / −52	0 / −25	−28 / −68	0 / −25	−77 / −117	120	140
0 / −25	+40 / 0	0 / −25	+12 / −28	0 / −25	−12 / −52	0 / −25	−28 / −68	0 / −25	−85 / −125	140	160
0 / −25	+40 / 0	0 / −25	+12 / −28	0 / −25	−12 / −52	0 / −25	−28 / −68	0 / −25	−93 / −133	160	180
0 / −29	+46 / 0	0 / −29	+13 / −33	0 / −29	−14 / −60	0 / −29	−33 / −79	0 / −29	−105 / −151	180	200
0 / −29	+46 / 0	0 / −29	+13 / −33	0 / −29	−14 / −60	0 / −29	−33 / −79	0 / −29	−113 / −159	200	225
0 / −29	+46 / 0	0 / −29	+13 / −33	0 / −29	−14 / −60	0 / −29	−33 / −79	0 / −29	−123 / −169	225	250
0 / −32	+52 / 0	0 / −32	+16 / −36	0 / −32	−14 / −66	0 / −32	−36 / −88	0 / −32	−138 / −190	250	280
0 / −32	+52 / 0	0 / −32	+16 / −36	0 / −32	−14 / −66	0 / −32	−36 / −88	0 / −32	−150 / −202	280	315
0 / −36	+57 / 0	0 / −36	+17 / −40	0 / −36	−16 / −73	0 / −36	−41 / −98	0 / −36	−169 / −226	315	355
0 / −36	+57 / 0	0 / −36	+17 / −40	0 / −36	−16 / −73	0 / −36	−41 / −98	0 / −36	−187 / −244	355	400
0 / −40	+63 / 0	0 / −40	+18 / −45	0 / −40	−17 / −80	0 / −40	−45 / −108	0 / −40	−209 / −272	400	450
0 / −40	+63 / 0	0 / −40	+18 / −45	0 / −40	−17 / −80	0 / −40	−45 / −108	0 / −40	−229 / −292	450	500

B.S. 4500 B